Best Practices
in Reengineering

Best Practices
in Reengineering

What Works and What Doesn't
in the Reengineering Process

David K. Carr

Henry J. Johansson

McGraw-Hill, Inc.

New York   San Francisco   Washington, D.C.   Auckland   Bogotá
Caracas   Lisbon   London   Madrid   Mexico City   Milan
Montreal   New Delhi   San Juan   Singapore
Sydney   Tokyo   Toronto

Library of Congress Cataloging-in-Publication Data

Carr, David K.
    Best practices in reengineering / what works and what doesn't in
the reengineering process / David K. Carr, Henry J. Johansson.
       p.   cm.
    Includes index.
    ISBN 0-07-011224-X
    1. Organizational change.   2. Corporate turnarounds.   3. Success in
business.   I. Johansson, Henry J.   II. Title.
HD58.8.C3632   1995
658.4'063—dc20   95-209
                                               CIP

  2 3 4 5 6 7 8 9 0 DOC/DOC 9 0 0 9 8 7 6 5

ISBN 0-07-011224-X

*The sponsoring editor for this book was James Bessent, the editing supervisor was Fred
Dahl, and the production supervisor was Donald Schmidt. It was set in New
Caledonia by Inkwell Publishing Services.*

Printed and bound by R. R. Donnelley & Sons Company.

McGraw-Hill books are available at special quantity discounts to use as premiums and
sales promotions, or for use in corporate training programs. For more information,
please write to the Director of Special Sales, McGraw-Hill, Inc., 11 West 19th Street,
New York, NY 10011. Or contact your local bookstore.

This book is printed on recycled, acid-free paper containing a minimum of
50% recycled de-inked fiber.

To Barb, Michael, and Brian for their patience, understanding,
and support.
David K. Carr

To Greta, who keeps me young and often reminds me to
stop and smell the roses.
Henry J. Johansson

CONTENTS

PREFACE

Our roots are in business operations improvement. We have been at it for almost twenty years. This includes early involvement in the application of Just-in-Time principles as well as many enterprise-wide TQM initiatives. These initiatives served many companies well under the mantra of continuous improvement.

We were delighted when business process reengineering began to emerge as a management tool in 1988; and, more importantly, with its broad-based acceptance which we track to 1991–1992. Our delight is traced to the professional satisfaction derived from the implementation of reengineered processes that yield dramatic gains in competitiveness. Previously focused on incremental gains, companies now adopted visions and objectives that made a step difference in their ability to compete through achievement of new levels of cost, service, cycle time, and quality performance.

As the experience base in business process reengineering grew, so did disillusionment. Surveys and articles often pointed to the high rate of failure and inability to realize the expectations set by new process design visions. At best, the reviews and surveys were mixed; failure as well as dramatic success stories.

With our delight clouded by reports of failure, we set out to identify "best practices" in business process reengineering. To identify those practices that are highly correlated with success, we researched 47 U.S.- and European-based companies that had achieved success or were far enough along in their efforts to forecast success. Basic research with these 47 companies was supplemented by a subset of more detailed case study reviews. The results of this research and the "best practices" are documented in this book.

There are many individuals who helped us analyze the research results and provided assistance during the preparation of this book. Robert Evanson, our longtime colleague from McGraw-Hill, was a major force at the outset. He pushed us hard and without this push we probably would not have started. Many of our colleagues from Coopers & Lybrand helped. Michael Bear and Amy Weber played a major role from start to finish in the development of the manuscript. They were the glue that held it together and deserve a major acknowledgment for their contribution. Matt Shinkman provided analysis and insights that made a difference. Wood Parker stepped forward and helped with case study materials and review insights that clearly added value. Mary Ann Fitzgerald and Karen Portman provided outstanding administrative support on a day-to-day basis. We also wish to thank the companies that agreed to be highlighted as case studies. These companies and their consultants provided detailed information about their projects, as well as access to executives and team members.

Jon Zonderman made it all happen. He listened to our ideas, received our diverse inputs, and molded our rambling discussions into a final product.

We hope that our research, documented herein, furthers the cause of reengineering and properly positions it as a management tool that can clearly make a difference in a company's quest for enhanced competitiveness.

DAVID K. CARR
Arlington, Virginia

HENRY J. JOHANSSON
Ft. Lee, New Jersey

Best Practices
in Reengineering

BEST PRACTICES IN BUSINESS PROCESS REENGINEERING

A major problem confronting business today is how to increase productivity, provide higher levels of service and responsiveness, and at the same time reduce costs. Since the beginning of the 1990s, companies have added time to market, response time, and customer service to a list of competitive "must have's" that already included quality, innovation, functionality, and cost.

Companies are increasingly coming to realize that traditional organizational structures, customer service philosophies, and business methods are no longer competitive in today's global market. They are also coming to realize that the old cost-cutting methods of slashing through departments and functions, and reducing head count willy-nilly, do not make them more competitive either.

What is needed is an organization that is *customer-focused* and *market-driven* in its external relations and *process-focused* and *team-oriented* in its internal operations. Only such an organization can look at the way work is performed across functions and seek to make those cross-functional operations more logical, effective, and efficient. Such an effort is at the heart of Business Process Reengineering.

THE ORIGINS OF BPR AND THE ONCOMING BACKLASH

Business Process Reengineering (BPR) is the business concept of the 1990s. But much of what is termed BPR is really little more than reworked Total Quality (TQ), continuous improvement, or systems-led implementation. In fact, some people even label as BPR efforts that are purely old-fashioned shortsighted cost cutting.

Consequently, as the second half of the decade begins, there is a gathering wave of discontent with and second-guessing over the broad concept of BPR. A host of studies are finding that many managers are unhappy with the results of their companies' BPR efforts. Much of that unhappiness, we believe, is due to companies' undertaking efforts with limited goals, less than full commitment, and poorly defined processes.

We believe that, when companies understand exactly what BPR entails before embarking on a reengineering effort—both the potential rewards and the real risks—and when leaders, senior executives, and managers all put their best effort into its implementation, companies are generally happy with what they accomplish.

We believe that, as of 1995, there is enough of a body of knowledge to say what works when engaging in Business Process Reengineering and what does not work. This is not to say that all BPR efforts can be shaped with the same cookie cutter. It is to say, however, that companies look to a set of "best practices" as guideposts in their own BPR efforts.

In this book we have sought to combine the methodology and guidelines we have been developing for five years with a survey of close to 50 U.S. and European companies, as well as a handful of in-depth case studies of companies who participated in the survey, to create this set of best practices.

BPR'S HISTORY, WHAT IT IS, AND WHAT IT ISN'T

The seeds of what we call Business Process Reengineering can be found in the Total Quality Management philosophies of Joseph Juran and W. Edwards Deming. Juran- and Deming-style TQM are clearly "process-focused," and they take a holistic view of workplace activities. Deming and Juran, both Americans, found little sympathy among Western companies for 30 years—until American and European companies began seeing the success that the Japanese had utilizing the philosophy of Total Quality.

Beginning in the late 1970s and into the 1980s, the philosophy was reintroduced into Western companies, starting in the manufacturing sector. Today these principles are firmly ensconced in most industries. Figure 1-1 is a timeline that puts the birth and development of TQM and BPR in the context of various management philosophies.

But this holistic, process focus flies in the face of more than half a century of Western management theory that has continually broken down work into ever more discrete and simple, repetitive tasks that can be performed by less and less skilled employees and, increasingly, by automation. These principles and theories generally are lumped together and talked about as Taylorism and the Scientific Theory of Management.

Many of us who were involved in TQM in the corporate world, or who have consulted with companies on TQM, worked hard to break through the entrenched Tayloristic and Scientific thinking and to get our companies or clients to think more in terms of processes, natural work groups, cross-functional work cells, and the like. We began moving companies toward a customer-focused, market-driven approach.

While TQM was beginning to push into Western management, another development was gathering momentum.

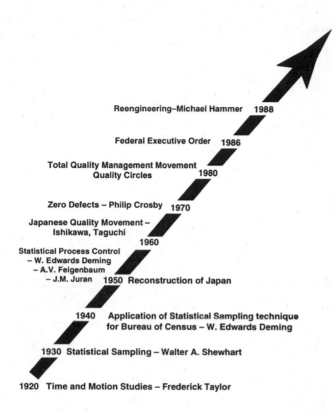

Figure 1-1. *TQM and BPR in context of changing management philosophies.*

Information Technology (IT) was the rage in the late 1970s and into the 1980s. IT departments were getting larger; companies were automating and adding sophisticated, expensive equipment.

In 1988, Michael Hammer, an IT consultant, wrote an article in the *Harvard Business Review* in which he exhorted readers to "obliterate, don't automate." IT and TQM needed to stop fighting, Hammer wrote. Many others of us were saying this privately in our companies or to our clients.

Hammer packaged the message perfectly and delivered it to an audience that was on the cusp of being ready to listen: Turning the cowpaths of most business processes into superhighways using the plethora of computer hardware simply doesn't work. The idea, he argued, is to tighten processes and to eliminate unnecessary and redundant steps. This was exactly what those of us who had been heavily involved in the TQM field had been saying—exactly what Juran and Deming had been saying all along.

Business leaders, many of whom were beginning to feel that their companies had ridden the Just-in-Time (JIT) and Total Quality waves as far as they could, were looking for something to help them navigate the rough waters they were beginning to feel as the Western economies started to lose steam.

By 1991, North America and Western Europe were in a serious economic slowdown, with some countries in recession. Western Europe was also looking at the need to bring its Eastern European neighbors, recently freed from the grips of central planning and repressive political regimes, into the world economy. Even Japan's economy was slowing.

At that time, the research and consulting firm IDC surveyed the U.S. business world and found growing acceptance for these ideas, and predicted huge growth in acceptance over the next three to five years in what was becoming known as Business Process Reengineering. Many companies had embraced the internal aspects of TQM—teaming, improvement, and the like. But narrowly focused TQM had a minimal impact on the corporate bottom line. Executives were now looking for business results, not just organizational results.

That prediction has been born out; by the beginning of 1994, the National Council for Manufacturing Sciences found that 80 percent of large manufacturers either were well into BPR efforts or were starting them up in 1994.

WILL THE REAL BPR PLEASE STAND UP?

Yet to this day, many still don't fully know what BPR is or what it is about.

In short, *BPR is about competitiveness.* That is the first driver, the company's competitive position. The "temple" of BPR, shown in Fig. 1-2, has "Competitiveness" written across the top. Its founda-

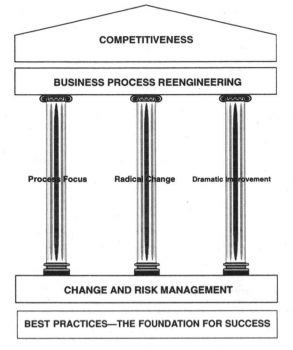

Figure 1-2. *The BPR "temple."*

tion is the best practices described in the this book, and its floor is firmly grounded in change management and risk management. The three pillars that the temple stands on are process focus, radical change (in behavior), and dramatic improvement in business results.

BPR is not about cost cutting. A reduction in cost, we have found time and again, is a natural occurrence in a well thought-out, well planned, and well executed BPR program. But if it is the sole reason a company goes into BPR, the effort will only result in costs being cut. The effort will not yield the kind of results we all know BPR can attain.

The first and foremost driver of BPR efforts is therefore the need to be competitive in the areas of cost, quality, lead time, delivery reliability, product characteristics, product support and service, and a host of other elements that the ever more sophisticated customer demands.

There are three key elements in our definition of BPR:

1. Process focus
2. Radical change
3. Dramatic improvement

Along with these three elements there is a need for constant and vigilant risk management.

PROCESS FOCUS

A *process* is a set of linked activities that take an input, transform it, and create an output. Ideally, the transformation that occurs in a process should add value to the input and create an output that is more useful to and effective for the recipient. Figure 1-3 shows what we mean by processes, and how they travel from the supplier to the customer.

To get the biggest bang for the BPR buck, the focus should be on *core business processes,* which directly touch customers, rather than on processes that are completely internal to your company. Many core business processes also directly touch suppliers. They are required for success in the industry sector in which the company does business, and the company's strategy should identify them as critical to excel at in order to match or beat the competition.

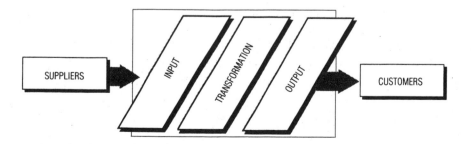

Figure 1-3. *How processes travel from the supplier to the customer.*

Figure 1-4 shows typical core business processes for companies in the manufacturing sector. For instance, new product development for an auto company or a first-tier auto supplier is clearly a core business process. Other industries have similar sets of core business processes. For example, order fulfillment is a core business process for an equipment manufacturer. Insurance policy underwriting is a core business process for an insurance company.

Typical Core Processes
Product & Process Design (Time to Market) Research & Development (Concept to Commercialization) Order Fulfillment (Order Sign to Delivery/Install) Conversion (Raw Material to Product) Procurement (Sourcing to Receipt) Logistics (Factory to Customer) Materials Management (Requirement to Consumption) Channel Management (Factory to Retailer) Supply Chain (Material Source to Customer)

Figure 1-4. *Typical core processes for manufacturing companies.*

It is possible to get good value from a BPR effort by reengineering processes that support core business processes. Yet, while these kinds of efforts may yield significant cost savings, they are really only marginal in terms of increasing a company's competitiveness. That's because they don't get at the second element of the definition, radical change. Yet often the effort in a BPR program is put into processes, rather than into radical change. That can lead to a lot of people being disheartened with the BPR results.

RADICAL CHANGE

Reengineering allows leaders, executives, and managers a rare luxury—the blank sheet of paper on which to reconstruct a commercial entity totally focused on today's and tomorrow's business problems. This should replace the organization that is too often engaged in looking backward at yesterday's business problems.

BPR does not seek radical change for the sake of radical change. The objective is competitiveness and, if possible, marketplace dominance. Radical change is a *characteristic* of this objective, an *outcome* of taking a process view and departing from the old way of doing business through functional departments. Figure 1-5 shows how processes differ from functions. Processes are horizontal while functions are vertical; processes cut across functions and functional activity feeds processes.

Although it is not necessary to destroy everything and start over, BPR allows people throughout the organization to move away from traditional ways of thinking and working. The radical change that comes about does not mean that a company must destroy all the assets in place; rather it provides a new way of leveraging a company's core competencies and meaningful management investments.

This, as we all know, is as frightening as it is exhilarating. And for many people it becomes a hurdle that is impossible to get over. In many instances the "cultural pushback" against the kinds of radical change necessary causes the BPR effort, as grand as it

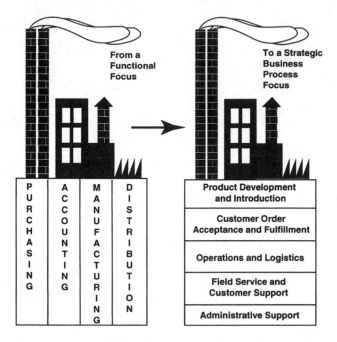

Figure 1-5. *Processes vs. functions.*

might be in concept, to fall flat in implementation. And again people become disillusioned.

DRAMATIC IMPROVEMENT

While large numbers of small, incremental improvement initiatives may be expected eventually to have a major cumulative effect, BPR expects dramatic improvement. Reengineering to achieve major improvements in performance takes place in the core business processes critical for competitive advantage. Dramatic "stretch targets" are set first; then the processes are reengineered to achieve those targets.

Again, if the rhetoric of leadership during the planning phase is that the BPR effort will achieve dramatic improvement, and if processes are then chosen for reengineering that are safe and easy, and if the BPR effort yields only modest or incremental improvement, people who put their time, effort, hearts, and souls into the effort will feel cheated.

Remember, BPR and the radical change it produces are risky endeavors. Leaders need to take the evidence presented to them by the business circumstances—and later by the business case presented by the planning teams—and decide if the potential rewards are worth the risks involved. They must lay out a detailed risk-management plan, and must be willing to stick with the effort over time if they feel the risk/reward ratio is worth going into a BPR effort in the first place.

WHY BPR?

BPR has emerged because it overcomes limitations inherent in other management concepts and approaches.

Downsizing deals with organizational structure, levels, and cost compression through the simple act of removing people. The work remains, as do the old methods, old systems, and old processes for getting the work done. Downsizing looks all right in the short term; it is only over time that the system begins to break down. Productivity can never increase to compensate for the people who have been removed from the processes. Therefore service to the customer suffers. Employee morale also suffers and burnout ensues.

Technology, particularly information technology, was promoted throughout the 1980s as a key to competitiveness. But a study by researchers at MIT concluded that more MIS does not mean more productivity or competitive distinction. Stephen Roach, the economist at Morgan Stanley, has noted often that, in the service sector where information technology investments are significant, productivity gains averaged only ½ to 1 percent per year throughout the 1980s and into the 1990s.

BPR views information technology as an enabler, not as a driver. Even when one can't envision radical change without quantum leaps in information technology, the change must be driven by a focus on changing customer-facing core business processes, not on changing information technology for the sake of changing the technology.

The relationship between BPR and IT is sorting itself out over time. Chief Information Officers and BPR leaders are less at odds. IT leadership is seeing how much easier major IT change is when it rides the wave of a strong and well-developed business case. And business leadership is seeing how much can be gained by the early involvement of IT and by the project management skills IT specialists bring to the BPR teams.

Functional Performance Improvement initiatives, such as manufacturing excellence through Just-in-Time, fall far short of the kind of radical change necessary. They are only one step in the value chain, and they fail to get out of "the box." These efforts take internal views of internal problems. This may lead to a situation in which the supply chain is suboptimized and customer value is affected.

In a similar way, *Materials Resource Planning (MRP)* focuses only on materials, and *Activity-Based Management* is focused solely on reducing cost through the activity-based costing (ABC) methodology.

Downsizing, a focus on technology, or functional performance management all suffer from another major failing; none of them is strategically focused. We adhere strongly to the philosophy that you can't get the process right until you *get the right process!* And you can't get the right process until you have a strong under-standing of the business strategy, which enables you to find the processes that do the most to implement that strategy.

Total Quality Management (TQM) has not delivered for many companies because, in too many instances, it has been narrow in focus and limited to improving the existing way of doing business. To be sure, this is not what Juran and Deming have advocated. Their notion of quality has been inherently linked to a company's strategy. But too often companies have taken the easy road and implemented quality programs that are not strategically based and are limited by being functionally oriented rather than cross-func-tionally process-oriented.

Many early practitioners and consultants in the quality movement used a model of training and education in their work. This often meant that a basic understanding of quality issues and the quality philosophy was instilled throughout companies through training workshops. But all too infrequently were employees asked to go back into their work processes and think about how to improve them horizontally across the process.

The incremental improvement that has been the goal of many quality efforts in Western companies does not work for a company that is not an industry leader or at least at the level of "best in class" with the competition. If you are behind, incremental improvements will too often keep you running as fast as you can just to stay in place.

As Paul O'Neill, Chairman of ALCOA, puts it:

> Continuous improvement is exactly the right idea if you are the world leader ... it is probably a disastrous idea if you are far behind in the world standard ... we need rapid, quantum-leap improvement. We cannot be satisfied to lay out a plan that will move us towards the existing world standard over some protracted period of time ... if we accept such a plan, we will never be the world leader.

Figure 1-6 shows some of the important differences between TQM and BPR.

TQM efforts have involved the workforce. Companies are learning how to do this well and are achieving payback. But, although there is a movement toward empowered teams, only a few companies have progressed to the point of mastering continuous improvement throughout the enterprise.

Despite this, TQM is a powerful experience, and we have found that many companies that get the best results from BPR have in place or have been through a rigorous TQM program. TQM helps companies create a common language, much of which involves problem description and problem solving. It also helps people learn how to work in teams on problem-solving efforts.

Factors	TQM	BPR
Type of Change	Evolutionary—a better way to compete	Revolutionary—a new way of doing business
Method	Adds value to existing processes	Challenges process fundamentals and their very existence
Scope	Encompasses whole organization	Focuses on core business processes
Role of Technology	Traditional support (e.g., MIS)	Use as enabler

Figure 1-6. *Fundamental differences between TQM and BPR.*

To be sure, TQM and BPR have a lot in common. Both are process-focused, emphasizing teams and shared values, and both use a similar tool kit for problem solving.

In practice, TQM tries to improve processes by identifying problems and solving them one at a time with continuous improvement as the goal. Often the processes are contained within functional boundaries. This is done across the entire company.

In BPR, the broadest possible look is taken when defining process boundaries. Functions are not allowed to constrain the view. By doing this, an enormous amount of waste can come out quickly. But once the processes are bounded, only a few are chosen for action.

Rather than attacking the many with incremental improvement, BPR attacks a few processes looking for radical improvement. The process after reengineering answers the question, How should the business be run? The question TQM answers is, How can each fragmented function be run better?

THE NEED FOR A NEW ORGANIZATION

Another driver for competitiveness in the 21st century is the recognition that what was good for the good old days is not good today. The hierarchical/functional organization addresses the need for command and control. This is an organization well suited for labor control and economies of scale. But functional barriers and handoffs make it slow. Too many levels within the organization institutionalize bureaucracy.

These characteristics of the functional hierarchy are not matched with the competitive needs of today's businesses such as flexibility, speed, time to market, and service and support. They are bureaucratic and slow, putting a premium on control instead of on responsiveness and entrepreneurial spirit. They are internally focused instead of being geared for the customer satisfaction.

Increasingly, businesses are realizing that this traditional Western type of organization needs to be broken apart and rebuilt in such a way that it is aligned with today's requirements for competitiveness.

HIGH LEVERAGE AND HIGH RISK

While it's easy to talk about the high degree of leverage a company can attain through a successful BPR effort, it's also necessary to understand, clearly and up front, that BPR is a high-risk venture. To be sure, many companies have had false starts or failures with BPR efforts.

Information Week's June 20, 1994 cover story was entitled "Reengineering Slip-Ups: Why Two of Three Efforts to Fix Corporations Fall Short." The article's first paragraph reads: "This year, American companies will spend an estimated $32 billion on business reengineering projects. Nearly two-thirds of those efforts will fail."

Without quibbling with how the magazine defines reengineering—far more broadly than we do—companies fail for some simple yet profound reasons.

One is an inability to meet the high expectations that BPR, by definition, sets for itself. Yet we would say that this is not true failure. In fact, most of the companies we surveyed—and they were chosen specifically because they were far along in the process and had shown some success—said they had failed to meet *all* of their expectations.

These companies, however, understood that the goals were developed specifically to stretch the organization and the individuals within it to their limits. *Stretch goals,* as we call them, are the key to making teams successful; these stretch goals force teams to think in different ways about how business should be done.

In some ways, BPR is like Xeno's paradox; if every step you take gets you half way to the goal, you will never actually reach the goal. But after the first giant step—the implementation of a reengineered process—you can settle into smaller steps—a continuous improvement of the reengineered process—and continue driving ever closer to your goals.

Other failures occur, we believe, because companies fail to truly appreciate the risk factors involved. They fail in areas of communications, measures, and accountability.

They don't put enough effort into rigorous change management. They don't communicate early, often, throughout the process, and with every employee in the organization. Executives don't "walk the talk." Either they don't show support early, or they let their support flag in the middle of the effort as their attention is drawn to "more important" matters, while the BPR work is being done in the trenches. They fail to communicate both their support of the effort and fail to be honest about the pain the effort will cause.

They try to do too much, reengineer too many processes. And they fail to set up performance measures that will help them track whether the effort is on course or not, and whether the reengineered process fulfills the objectives set for it.

Some companies fail to put in place good project leaders with project management skills at the reengineering implementation level. Or they don't give these project managers enough autonomy. On the flip side, some don't put enough accountability on these project managers.

How Our Study Fits with Others

Reengineering in its broadest sense has been the subject of a number of studies beginning in 1993, and in some instances the reporting of those studies has muddied the waters further as to what actually constitutes BPR. Arthur D. Little surveyed executives at 350 companies for its study "Managing Organizational Change"—not specifically a study of BPR. Yet *Information Week* said the study was of 350 companies involved in reengineering.

Arthur D. Little found that only 16 percent were fully satisfied with their change efforts, while 45 percent were partially satisfied and 39 percent dissatisfied. Also, 68 percent reported that their efforts to change the organization had created unintended problems.

This problem with defining BPR exactly was born out in a study by Forrester Research, a small Cambridge, Massachusetts company that surveyed 50 companies claiming to be reengineering business processes. Forrester found that 42 percent of its survey population was really engaged in efforts that would lead to incremental changes, not radical change. Another 28 percent really weren't reengineering at all. Only 30 percent were undertaking what we would call true BPR.

By far the largest study of BPR was undertaken by CSC Index, the organization which Michael Hammer associated with for years and which is still often seen as the birthplace of BPR. That study surveyed 621 companies—497 in the United States and 124 in Europe—and found that 69 percent of American companies and 75 percent of European companies surveyed were, in early 1994, reengineering processes, and the rest were in planning or seriously considering a reengineering effort. David Robinson, CSC Index's president, told *The Wall Street Journal* that his company's study also found confusion over the definition, saying "people tend to call anything that moves reengineering."

CSC reported that:

> North American reengineering activity is feverish in labor- and capital-intensive industries that need to get lean [read "cut costs"] fast: automotive, telecommunications, aerospace, chemicals, pharmaceuticals, heavy manufacturing and consumer products.

The CSC survey (done by mail, with only five in-depth telephone interviews) does a good job of pointing to the differences between North American and European efforts. Most North American reengineering efforts are aimed at points of direct customer contact: 25 percent in the cutomer service process; another 16 percent in order fulfillment; and another 11 percent in "customer acquisition" activities, such as sales and marketing. In the companies surveyed by CSC, very little reengineering is going on within the "guts" of manufacturing operations or internal processes in service organizations.

By contrast, CSC found that in Europe 23 percent of reengineering activity is taking place in manufacturing processes or internal processes in service companies, with 15 percent in customer service and another 15 percent in distribution, and a further 13 percent in order fulfillment.

Although we did not structure our questions the same way, our results show much the same trend; reengineering in the United States is occurring to a great extent in customer-focused areas where communication with the customer is vital and information technology can often have a dramatic effect.

The CSC Index study also found about a 25-percent failure rate, far lower than many anecdotal estimates, which had ranged in the press as high as 85 percent—the number used by *Information Week* in its cover story the week before the CSC Index study was made public.

From its study, CSC Index derived three basic rules:

1. Have strong, rigorous project management.
2. Have unhesitating support from executives.
3. Set aggressive goals.

We couldn't agree more. Those three rules key in very closely with many of our best practices. We also have a number of other best practices. And that was our goal in conducting our study, to identify practices in carrying out a rigorous reengineering effort that correlate with success.

We wanted to get beyond the quibbling over definition. We also wanted to get beyond the quibbling over just exactly how you define success, and whether companies were or were not successful by anyone's measure except their own.

A STUDY OF SUCCESS

It was easy to identify 150 potential candidates for the study. After a brief telephone discussion, our researchers from Dataquest (a subsidiary of Dun & Bradstreet) were able to determine if the candidate company should participate in the full 90-minute telephone interview. Criteria for determining this were:

Is the company far enough along in the reengineering process?

Is the company really reengineering and not just seeking incremental improvement?

Does the company consider the effort (as it stood at the time the initial interview took place) *successful enough* to spend 90 minutes discussing what steps the company has taken to guarantee a level of success it felt comfortable with?

We found 47 companies that fit the bill. These are companies that have set out their goals and reached them to a great degree. Of the 47, 34 percent were totally satisfied with their efforts and another 50 percent were highly satisfied. These results are shown in Fig. 1-7.

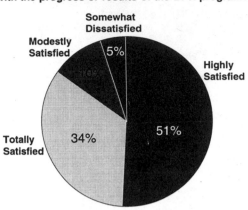

How would you rate your overall level of satisfaction
with the progress or results of the BPR program?

Somewhat
Dissatisfied

Modestly
Satisfied — 5%

Highly
Satisfied

Totally
Satisfied — 34%   51%

Figure 1-7. *Level of satisfaction.*

What's more important, 95 percent said they would be willing
to undertake the project again, despite difficult and often painful
experiences in the effort. The reward was worth the risk, the gain
worth the pain.

Even if companies have not enjoyed what might be called "textbook" success, all feel they have enjoyed some success, even if it is just
the intangible success of having an organization that is more mobilized
to meet the challenges of today's business world because employees
understand the concepts of processes, team organization, and teamwork. And, in the end, that may prove more important to some companies than textbook success. For in today's rapidly changing business world, success in any particular process may not fully translate
into long-term business success, as might the ability to define, engi-
neer, and implement a new or better process when necessary.

We found a study population that clearly shows a commitment
to rigorous organizational change and to the BPR concept itself.
Over half of our respondents had prior experience with a change
effort—mostly TQM. More than half were in the implementation
part of the reengineering effort. The average length of their BPR
efforts was two years and four months, and about 30 percent of
our respondents said they believe, "BPR is never really over."

Of those who had been involved in a previous change effort:

53 percent said BPR represented a "more radical change" than they had previously undergone.

47 percent said the change was "more complex."

26 percent said the change had a larger scope.

21 percent said it was the first time they had really undertaken cross-functional change.

We believe most of our survey respondents went into their BPR efforts for the right reasons: 51 percent to improve customer service, 49 percent to reduce cycle time, 37 percent to reduce costs, and 26 percent to improve quality. The full range of reasons for embarking on BPR is shown in Fig. 1-8.

Not all of our respondents had or would tell us about "quantifiable" goals for their BPR efforts. But those who did report them expected dramatic gains. For example, one respondent's goals were a 25-percent increase in customer satisfaction ratings, a 30-percent decrease in costs, and a 40-percent decrease in cycle time and in-process defect rate from 6000 ppm to 40 ppm. That's ambitious!

Eighty-nine percent of our respondents felt their goals were realistic. Statistically, with a sample of only 47 companies, it was not possible to tell if having realistic goals correlates with overall satisfaction in results. We did find anecdotal evidence that having stretch goals helps achieve enhanced results, and the question certainly bears further scrutiny.

Setting goals leads us to a key best practice.

BEST PRACTICE: LINK GOALS TO CORPORATE STRATEGY

In all cases, our survey respondents linked their goals to their corporate strategy. Goals were company-specific, but tended to be either "customer-focused" or "quality-focused." Some examples:

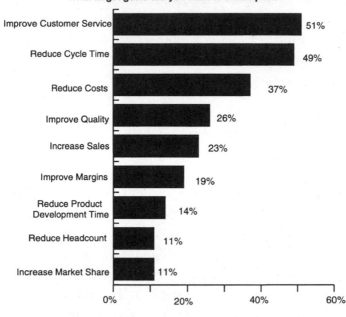

What target goals did you want to accomplish?

Goal	Percentage
Improve Customer Service	51%
Reduce Cycle Time	49%
Reduce Costs	37%
Improve Quality	26%
Increase Sales	23%
Improve Margins	19%
Reduce Product Development Time	14%
Reduce Headcount	11%
Increase Market Share	11%

Note: Multiple responses permitted.

Figure 1-8. *Reasons for embarking on BPR.*

Our strategy is to improve service while decreasing costs.

Our corporate strategy is to be the number one provider of electric utility service.

Our goal is to focus on quality and service, while reducing all nonvalue-adding activities.

We'll discuss this best practice in detail in Chap. 4 when we talk about the planning stage of BPR, when you endeavor to "do the right process" in preparation for doing the process right.

THE OTHER BEST PRACTICES

As we went through the survey results, we found 16 additional best practices that correlate with results. (Appendix A is a complete listing of best practices. Appendix B is a list of companies that participated in the full 90-minute interview. Appendix C is the interview itself.) We didn't always find that companies had followed these best practices; but, when companies did not, we often found that they wished they had. Some companies were very explicit about lessons they learned early in their BPR effort that allowed them to become far more successful over time. For example, maybe only 25 companies engaged in a particular best practice, such as linking goals to strategy—slightly over 50 percent of the survey respondents. You might ask, "If only 55 percent did it, how could it be a best practice?" Suppose that, in response to the open-ended question, "What could you have done better?" 18 *other* respondents said, "We should have linked our goals to strategy more clearly." You now have a total of 43 companies—over 90 percent of respondents—responding positively to the best practice.

We've tried to organize our chapters by combining a natural group of best practices with our particular consulting framework.

In Chap. 2, we discuss issues of change mangement, the human dimension of BPR. Too often it is tempting to think of BPR as merely a rethinking of the way work processes are performed. But, since work is performed by people and not by machines, when you think about changing work processes you naturally have to think about the impact these changes will have on people. BPR is a massive change effort (even though not all massive change efforts are BPR). For this chapter, we found four best practices.

1. *Best Practice: Recognize and articulate an "extremely compelling" need to change.*
2. *Best Practice: Start with and maintain executive-level support.*
3. *Best Practice: Understand the organization's "readiness to change."*
4. *Best Practice: Communicate effectively to create buy-in. Then communicate more.*

In this chapter, as in all chapters, we will also present "prospective best practices." These are issues we have seen in our consulting experience to be important enough that we would consider them best practices, even though they did not come out that way through our survey. In some instances, it may be a practice or issue that hasn't bubbled to the surface yet or hasn't been embedded in the consciousness of those involved in BPR efforts. It might be seen as an "extra credit," something that gives a BPR effort that is hitting all the targets a little extra push.

For this chapter, the prospective best practices are:

Instill in the organization a "readiness and commitment" to sustained change.

Stay actively involved.

In Chap. 3, we discuss the *prework,* or organizing for a BPR effort. These organizational logistics are extremely important, for if the proper structures aren't put into place within the day-to-day operation of the BPR effort, it will not obtain the results it should. For Chap. 3, there are three best practices:

5. *Best Practice: Create top-notch teams.*

6. *Best Practice: Use a structured framework.*

7. *Best Practice: Use consultants effectively.*

The prospective best practice in this chapter is relatively simple:

Pay attention to what has worked.

Lessons learned are always important. And most modern companies have tried many management tools to enhance performance and competitiveness. There have been successes, and BPR should build on those successes.

In Chap. 4, we begin the real work of BPR. This is where you strive to do the right process in preparation for doing the process right. This is the point at which corporate leadership needs to create the vision of "where we should be." At the same time, the BPR teams are determining "where we are now." The gap between those two perspectives is the distance that the BPR effort must travel. The team also studies the company's own process capabilities, and builds a business case for closing the gap by reengineering particular processes where the current capabilities will allow for closing the gap.

We have identified four best practices for this chapter:

8. *Best Practice: Link goals to corporate strategy.*

9. *Best Practice: Listen to the "voice of the customer."*

10. *Best Practice: Select the right processes for reengineering.*

11. *Best Practice: Maintain focus: Don't try to reengineer too many processes.*

Our prospective best practice in this chapter is:

Create an explicit vision of each process to be reengineered.

We believe this is an area that many companies are not fully exploring in their efforts. Clearly, the effort to understand "where we are" and to analyze the gaps between "where we are" and "where we want to be" is a daunting job. But the gap cannot be closed in the most effective manner unless you create an explicit process vision.

In Chap. 5, we get at the issues involved in actually reengineering the processes chosen in the planning stage. At this stage of the effort, team membership will often change as some high-level people, those involved in creating the vision of the future state and in analyzing gaps between current and future, leave the team. They are replaced by people with more detailed, expert knowledge in the processes chosen for reengineering.

It's important that teams maintain their momentum, both in an organizational sense and in the sense of not reacquiring the previous group's understanding of the as-is and future states. We have identified three best practices in this arena:

12. *Best Practice: Maintain teams as the key vehicle for change.*
13. *Best Practice: Quickly come to an as-is understanding of the processes to be reengineered.*
14. *Best Practice: Choose and use the right metrics.*

As our prospective best practice, we believe teams at this stage need to:

Create an environment conducive to creativity and innovation.

Breakthroughs often happen by serendipity. But there are tools that help people organize their thinking as individuals and as teams so that they can bring about an atmosphere in which innovation and creativity are more likely to happen. We have found that clients who put the time and effort into having their process-reengineering teams go through these exercises have been richly rewarded.

A second prospective best practice is to:

Take advantage of modeling and simulation tools.

Process modeling and simulation can be as simple as process flow diagrams or as sophisticated as allowed by complex computer simulation programming. Different companies will feel comfortable with varying levels of complexity along the continuum. But it is important in the rigorous BPR reengineering stage to engage in the most sophisticated modeling your organizational culture will tolerate.

In Chap. 6, we get to the point of implementing the reengineered processes. In this chapter, we identify two final best practices:

15. *Best Practice: Understand the risks and develop contingency plans.*

16. *Best Practice: Have plans for continuous improvement.*

In addition, the prospective best practice revolves around the need to:

Align the infrastructure.

By this we mean institutionalizing and embedding in the fabric of the corporate culture the kinds of human relations changes that are necessary to sustain the reengineering effort. Leaders, executives, managers, facilitators, and others can't go back to the old ways of working once the BPR effort is over and the change management concerns are done with.

The reason for this is simple. As so many of our survey respondents said, BPR is never really over.

The second prospective best practice is to:

Position IT as an enabler, even if the extent of the IT change necessary is great.

Although IT changes can often enhance the value of the BPR effort, IT should never drive the changes without a clear business case for change. This can cause you to fall into the trap of "paving cowpaths." Even if the IT potential drives the initial thinking about change, it is necessary to step back from the technology, understand the current process, and reengineer it before automating it or before adding or enhancing present information systems.

In Chap. 7, we will tell you the story of one company that has made continuous reengineering and change the watchword of its strategy in the 1990s and into the 21st century. We will present our thoughts on how you can create within your company an "improvement-driven organization."

THE HUMAN FACTOR: BPR AS A CHANGE MANAGEMENT EFFORT

Best Practice 1: Recognize and articulate an "extremely compelling" need to change.

Best Practice 2: Start with and maintain executive-level support.

Best Practice 3: Understand the organization's "readiness to change."

Best Practice 4: Communicate effectively to create buy-in. Then communicate more.

Prospective Best Practice: Instill in the organization a "readiness and commitment" to sustained change.

Prospective Best Practice: Stay actively involved.

Business Process Reengineering is a two-pronged effort.

One portion of the BPR undertaking might be termed *technical*. This part involves the identification of:

Processes throughout the business.

The core business processes that drive the company value.

The subsequent reengineering of one or more of those processes in order to tighten connections with customers, streamline operations, and eliminate wasteful, nonvalue-added steps in the identified processes.

The other portion of the BPR effort might be termed *behavioral*. This component involves the identification of changes in the "way people work" throughout the organization that will have to take place in order for the technical aspects of BPR to be successful, and the subsequent management of those changes.

Despite all that has been written on change management and organizational development in the last two decades or more,

American corporate culture is still likely to seek solutions to business problems by working on the technical side of the equation. Not enough thought or effort is put into the behavioral part of many business change efforts. As Tom Terez, a management strategist and leader in the effort to change companies, put it, "The American corporate landscape is littered with the remains of technically sound programs that have been crushed by employee resistance to change."

Part of the reason for this is that many companies, as part of their previous efforts to alter the business, have failed to build a coherent, unified corporate culture with a unity of purpose. Too often, these companies have a hierarchical structure that leaves people feeling powerless and defensive, at the whim of supervisors or managers just ahead of them and senior managers.

In the United States and Europe since the mid-1980s, there has been a growing movement away from hierarchical structure, toward an empowered employee base, often working in natural work teams and even in self-managed teams. Just about every truly successful company today has broken the old codes; they have engaged every single worker in the improvement process.

A wonderful example of an industry that has changed the way it seeks change is the U.S. automotive industry. Since the days of Henry Ford and the River Rouge plant, the automotive industry was hierarchical and autocratic—some would say tyrannical. Efforts to change the industry in the 1970s and early 1980s in response to the influx of Japanese cars were first aimed at technical solutions: automation, efficiency, better engineering. However, only when the industry turned to process-oriented, team-driven tools, techniques and philosophies—Just-in-Time manufacturing and Total Quality later in the 1980s, and intensive reengineering in the early 1990s—did the industry experience a real turnaround.

Many companies across the industry spectrum did this through aggressive TQM efforts. BPR extends these efforts. That is one reason we say that companies that have engaged the TQM movement over the past decade have a leg up when engaging BPR.

BPR can and should be used to extend the gains in employee empowerment and teamwork created under any TQM effort. TQM efforts are sometimes limited not only because they seek incremental rather than radical improvement, but because they then make those improvements within the old-fashioned functional framework that has typified Western companies since the early part of this century. By working within the existing functional framework, you are all but resigned to small gains.

Many TQM efforts fail to break down the walls between departments and functions that often hinder change. Despite the rhetoric about valuing teams and empowered employees, too many companies continue to have in place systems, procedures, and practices that focus on individual effort and top-down control, usually within a functional context. They continue to reward individual compliance with orders from above; while the rhetoric is about teams and processes, the metrics too often are still tied to individual and functional success. The resulting conflicting messages contribute to widespread cynicism and distrust of management motives.

Attempts to integrate departments and improve customer service are often thwarted by department managers who see only that they will lose power through integration. As a consequence, departmental processes are more powerful than business processes. But without strong business processes, people waste their efforts by duplicating activities or doing things that offer no payoff to the organization.

Fear of failure, fueled by a management style that punishes mistakes and that does not recognize anything short of spectacular success, leads to hiding poor performance rather than regarding it as an opportunity to improve. Fear of failure prevents people from changing their routines and rituals. Autocratic, coercive management is one of the root causes of a high resistance to change.

A MODEL FOR CHANGE

In our efforts to help companies manage change, we use a model designed by W. Warner Burke and George Litwin, shown in Fig. 2-1. The model provides a framework for understanding the structure of organizations and is deliberately hierarchical; changes made at the top of the model carry more "weight" in the organization than those made at the bottom. As Burke and Litwin write in their 1992 *Journal of Management* article:

> [O]rganizational change, especially an overhaul of the company business strategy, leadership and culture have more "weight"

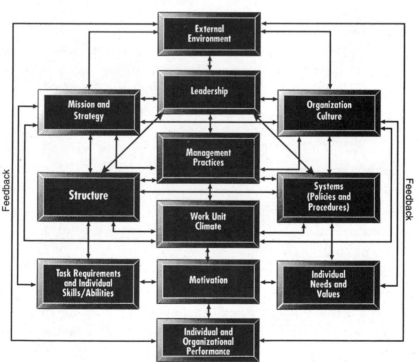

Figure 2-1.   *The Burke-Litwin Model.*

than structure, management practices and systems: that is, having organizational leaders communicate the new strategy is not sufficient for effective change. Culture change must be planned as well and aligned with strategy and leader behavior.

While changing the variables that run horizontally along the top of the model changes the entire business system, changing the variables at the next level down—structure, management practices, and systems—does not necessarily change the entire system. Burke and Litwin describe this as the difference between transformational and transactional change, leadership, and management.

They argue that leaders create *transformational change*— change that is deep rooted, pervasive, and long lasting. Managers, on the other hand, create *transactional change*—more short-term changes, based on transactions between people or groups who essentially agree to do something for one another in return for the other party's doing something.

Climate changes are changes in what Burke and Litwin call the "collective current impressions, expectations and feelings that members of local work units have that, in turn, affect their relations with their boss, with one another and with other units." Climate changes can be driven by managers. But *changes in culture*—the "collection of overt and covert rules, values and principles that are enduring and guide organizational behavior"—can be driven only by leaders.

This is where we see one of the clearest distinctions between Total Quality, as practiced by Western companies, and Business Process Reengineering. TQ efforts are often left in the hands of middle managers, who are asked to create transactional changes across every dimension of a business in order to achieve continuous, incremental improvement. BPR must be driven from the top in order to create transformational change.

As illustrated at the top of the Burke-Litwin model, change within an organization is spurred by the external environment; hence our first best practice.

BEST PRACTICE 1: RECOGNIZE AND ARTICULATE AN "EXTREMELY COMPELLING" NEED TO CHANGE.

With BPR the compelling need is driven by the marketplace and the competitive environment. Without a compelling need to increase competitiveness, efforts to transform a company will run up against the "who cares" syndrome.

To control and shape the direction of change, a company must develop a thorough understanding of the desired state—what executives want the company to be like in three to five years—and of the current state. This means that individuals within the organization must change, as well as the organization itself. The future state may not be the end state, but an "improved" state. Between the as-is and the desired state, the organization will go through a *transition state*; this is illustrated in Fig. 2-2.

The size of the gap between the desired and current states argues for the compelling need to change. In short, the pain of remaining in the current state has to be worse than the pain involved in trying to make change occur.

In our survey, 71 percent of respondents recognized an "extremely compelling" need to change. Only 10 percent felt the reasons being articulated for the changes sought by senior executives were less than compelling. Some of these compelling needs included:

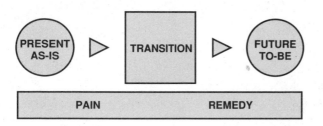

Figure 2-2. *The transition from the as-is state to the future state.*

A U.S. electronics company saw a "loss of market share beginning in the late 1970s," and had customers who "feel that our company's quality stinks."

A U.K. public utility that recognized, "Customer complaints were at an all-time high in 1990, and with impending privatization the customer service part was deemed to be absolutely critical to improve."

Another U.K. utility, facing privatization, discovered that, "The privatization program required a total rethink of the way in which business was to be conducted."

A Canadian government department realized that, "Until 1988, our process had been done the same way for over 100 years."

A European auto manufacturer found that, "After the failed merger, [we] realized that to survive we needed to become intensely competitive, and driving costs down was a major part of it."

As Burke and Litwin suggest, the compelling need to change comes from an external stimulus. Sixty-four percent said the forces triggering the need to change and the BPR effort were competitive pressures, while 16 percent said the forces were regulatory and 16 percent said the forces were a significant market opportunity.

Only 7 percent said poor customer satisfaction was a triggering force, although 51 percent had said that improving customer services was a major goal of the BPR effort. The entire array of competitive forces impinging on survey participants is shown in Fig. 2-3.

Clearly, best-in-class levels of customer service is a competitive issue today, and companies cannot wait for poor customer satisfaction to trigger a change effort. It may simply be too late.

Sometimes the current state makes itself known with unusual bluntness and force. Xerox, Motorola, and Ford are just three of the multitude of companies for whom a compelling need to change was obvious in the 1980s—to executives, to Wall Street

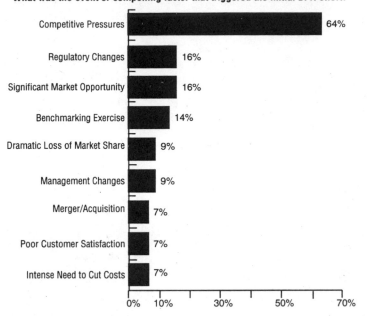

What was the event or compelling factor that triggered the initial BPR effort?

Factor	Percentage
Competitive Pressures	64%
Regulatory Changes	16%
Significant Market Opportunity	16%
Benchmarking Exercise	14%
Dramatic Loss of Market Share	9%
Management Changes	9%
Merger/Acquisition	7%
Poor Customer Satisfaction	7%
Intense Need to Cut Costs	7%

Note: Multiple responses permitted.

Figure 2-3. *The range of competitive forces affecting survey participants.*

analysts, and to the public. These three companies, like many others, practically reinvented themselves with agressive TQM and BPR efforts in the late 1980s and into the 1990s.

But for every Xerox, Motorola, or Ford, the corporate landscape is littered with the hulks of companies that either failed to recognize the compelling need or that were too timid to try to reinvent their companies to face these challenges.

A SUCCESS STORY

Lawrence Bossidy became CEO of AlliedSignal, Inc. in 1991. A company forecast right at the start of his tenure predicted negative cash flow for both 1991 and 1992. Debt was 42 percent of capital, and surveys showed that executive morale was lower than the company's bottom line.

Clearly, change was needed. And it was clear that leadership was required to create a vision and to direct the change. Bossidy, who had left a job as Jack Welch's right-hand man at GE to take on AlliedSignal, was hailed in 1993 by *Fortune* magazine as one of the few "masters of corporate revolution." (Bossidy put his personal stamp on the company even to the extent of joining the words Allied and Signal together into one word—AlliedSignal—because he felt the space between the two words was wasted effort for a reader.)

"We went into 1992 with three objectives," Bossidy says. "To make our numbers; to make Total Quality a reality, not just a slogan; and to make AlliedSignal a unified company."

(A full discussion of AlliedSignal's success in Business Process Reengineering appears in Chap. 7.)

Bossidy's use of the phrase Total Quality is a shrewd stroke of leadership. What he did was engage Business Process Reengineering with a zeal few other corporate leaders have. But, since AlliedSignal was already involved with a TQ effort—albeit with mixed results to that point—Bossidy retained the nomenclature in an effort to enhance morale around subsequent results and to try to head off the resistance that often occurs when employees feel that executives are engaging in the "change program of the month." Maintaining this focus, not flitting from one change fad to another, helps employees to understand that there really is commitment from the top to set a course and stick with it.

Bossidy used the report he received on his first day as AlliedSignal CEO as the key example of the compelling need to change. Along with this, he set as the desired future state a set of aggressive goals for performance on a variety of measures:

6 percent annual gains in productivity

Increasing operating profits from 4.7 percent in 1991 to 9 percent in 1994

Increasing working capital turnover from 4.2 times per year to 5.2 times per year

Raising return on equity from 10.5 percent to 18 percent by 1994

Bossidy also worked with AlliedSignal's top 12 executives to develop the company's "to-be" vision, which commits the company to "strive to be the best in the world." The vision specifies a customer focus, teamwork, innovation, and speed as key means to the end. The vison also specifies that AlliedSignal will become a "Total Quality company by continuously improving all our work proceses to satisfy our internal and exernal customers."

AlliedSignal's definition of a total quality company includes:

1. Producing satisfied customers.
2. Focusing on continuous improvement.
3. Having highly motivated and well-trained employees.

It also includes taking high-quality steps to achieve excellent business results, such as managing by fact, taking a process orientation, and using a business planning process that has quality goals, steps to achieve them, and the means to measure them.

FINDING A COMPELLING NEED

The most difficult thing for a company to do is to find a compelling need during good times. It's easy for a Xerox to see the compelling need when its core photocopying business is being cut to pieces by competitors, or for Ford to see the compelling need when its cars are rated low by consumers in quality. Such compelling need is often called a *burning platform,* as depicted in Fig. 2-4.

The way to establish a compelling need, in the words of Jack Welch, CEO of General Electric, is for the leader to "change before you have to." After that, you have to be willing to hold the line, take a hit if necessary in terms of financial performance and/or public perception, and see the change through.

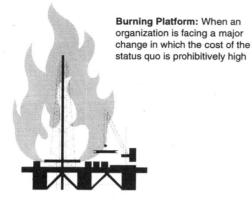

Burning Platform: When an organization is facing a major change in which the cost of the status quo is prohibitively high

Figure 2-4. *A "burning platform."*

For instance, it took real forward thinking for Salomon Brothers to reengineer its back office functions and move them from New York City to Tampa in the early 1990s.

It is much harder for an electric utility company to see the compelling need when it still shows a healthy return on investment and still works in an environment where it is a monopoly, despite all the talk about competition and deregulation.

At Aetna, impetus for the Life Claims Project (discussed in detail in Chap. 5) was "only moderately compelling" in and of itself. However, "it was part of the overall effort to change the corporate culture to one that is customer focused, and that requirement for change was extremely compelling."

One way to make a case for a compelling need in good times is to continuously benchmark your efforts against the "best" organizations, regardless of what industry they are in and whether they compete with you directly or not.

As the case of AlliedSignal so vividly points out, the compelling need to change is driven by market forces and articulated through the mission and corporate strategy. The agenda for change is determined by understanding what has to change and what has to be developed within an organization in order to implement the leaders' vision—the corporate mission and strategy. The key point is that there must be a powerful strategic argument for cultural change.

Cultural change is a behavioral response to strategic objectives: What do we have to do differently to achieve the corporate mission and strategy? Or conversely, What do we do so well that it should inform our strategy?

Senior executives must take the leadership vision and drive it down into the organization through defining and widely communicating the behavioral changes that are required. While they say that the external environment sets the stage for organizational change, only in the hands of a good leader can the stimulus from the external environment be fully harnessed and used as a catalyst for internal change. This brings us to our second best practice.

BEST PRACTICE 2: START WITH AND MAINTAIN EXECUTIVE-LEVEL SUPPORT.

Corporate executives cannot be change leaders until they have commited to change themselves. This is more than just analyzing the company's needs and deciding it must change.

Once a top executive has intellectually accepted the need for change, his or her heart and soul must follow. And then the leader must grab the heart, soul, and intellect of every member of the executive team.

For too long, many top executives have ignored their role of leader in favor of one as top manager. Instead of creating a broad vision and direction for the company, they have sought to manage every step a company makes, often losing sight of the vision for quarterly financial results. But BPR has as its central tenet that nothing is more important for the business today than for executives to think about tomorrow.

Figure 2-5 shows the different characteristics of leaders and managers.

Leaders	Managers
Establish and communicate the vision	Carry out the vision
Motivate employees	Motivate, guide, and direct employees
Establish and exhibit fundamental values	Translate fundamental values into business results
Focus on the future	Understand future direction but monitor the present

Figure 2-5. *Characteristics of leaders and managers.*

The Harvard Business School Professor John Kotter, in his 1990 book, *A Force for Change: How Leadership Differs from Management,* says that producing change is the primary function of leadership. While managers plan deductively and produce orderly results, leaders set direction and develop vision and strategy. Managers organize and staff activities, while leaders align people by communicating the vision and empowering action.

Jack Welch told *Fortune* magazine in 1989, "The job of the leader is to take available resources—human and financial—and allocate them rigorously ... and to articulate the vision to employees."

In 1993, Marc Sternfeld, Managing Director of U.S. operations for Salomon Brothers, Inc., told a team of researchers for the *Harvard Business Review,* "I personally have to be the lighthouse to make sure we're moving toward the team-culture direction.... I'm the one who doesn't move off the vision principles.... I have to make sure we keep moving in the right direction...."

If top management loses focus on the BPR effort, they will lose support for it and commitment to it, and every member of the organization will see this loss of focus, support, and commitment. Throughout the BPR effort, the wavering of executive support and commitment could be a sign that you have chosen the wrong process(es) to reengineer.

Leaders depend on their senior management team to carry the vision and direction to the next level of day-to-day organization and operations. Howard Sperlich, a top member of Lee Iacocca's management team at Chrysler in the early 1980s, told Warner Burke in 1984:

> The dramatic leadership that Iacocca provided was in the classic leadership mode—a guy you believe in, you'd follow into battle. He was so strong that, at the worst time, knowing his hand was on the helm, a dealer would keep his money in the business; a supplier would bankroll us ... a guy would come over from Ford.... He provides leadership in battle, his fundamentals are right, and he behaves consistently.... My own leadership style is a little bit like his, but on a smaller scale. I'm not big enough, and I'm not supposed to be.... I'm more actively involved in trying to promote results through common goals and an enabling style.... But as chairman and CEO, he's got to establish the fundamental values.

Ronald Compton, now chairman of Aetna, has been pushing for major changes at the giant insurance company for years. Compton has been quoted numerous times as saying that, now that he is chairman, he will change the corporate culture of the old-line, paternalistic company himself if he has to.

Because of the dramatic impact BPR has on the organization and its business, leaders cannot "delegate" BPR to the group of managers and executives who report directly to them. BPR demands more day-to-day involvement on the part of senior executives than any other effort, including TQM, which leaders can "kick off" and then delegate and leverage.

Change is painful. The necessary changes may end careers, including that of the change leader. While top corporate managers can rely on peers or on the CEO to help them understand the need for change, for the top executive and change leaders, that understanding must come from within.

Jim Eibel of Ameritech, who is leading the change from a company that enjoys monopoly protection to one that will live in a competitive environment, says:

> I know that once we've positioned ourselves as a competitive company this job—my job—won't exist any more. I've had some sleepless nights.... It's important that you really do believe in it, that you're fully committed to the necessity for change.

LEADERS: BORN OR MADE?

Many of our examples cite well-known corporate leaders. But what of other organizations?

The management theorist Abraham Zalesnick has argued that leaders are born and not made. To some extent this is true, and businesses that allow for natural leaders to reach the top of the organization are richly rewarded. But in most businesses, culture has caused managers to reach the top. Therefore, to function in the real world, companies must ask whether leadership skills may be obtained later in life to a sufficient degree to allow top executives to lead their companies through massive change efforts like BPR.

We think leadership can be taught. We also believe that more and more corporations are learning the value of leadership as opposed to management, and that they are grooming better managers for leadership positions by sending them for special leadership training, which is increasingly being offered in midcareer business school programs and in special training environments.

The best example of an organization that has long believed in the "making of leaders" is the military. From very early in their careers, military officers are put through increasingly rigorous and intense leadership exercises. That may be why it has long been the case for senior military officers to retire after long careers and immediately step into leadership roles in businesses in a wide range of industries.

Along with "growing" leaders, it is important for companies to put in place structures, management systems, and policies and procedures—the second tier of the Burke-Litwin Model—that allow senior executives the time and energy to carry out their leadership duties. Leadership is time-consuming and energy-consuming, and, if top executives are busy fighting fires, they don't have time to exercise leadership—whether they are natural leaders or learned leaders. Structures, systems, policies, and procedures that drive much of the day-to-day management decision making into the hands of managers below the senior-executive level leave senior executives the time and energy they need to do the things leaders have to do.

And one of those "things" is what our third best practice is all about.

BEST PRACTICE 3: UNDERSTAND THE ORGANIZATION'S "READINESS TO CHANGE."

One way of looking at this issue is that, if the leader has effectively shown the compelling need to change, there should be no doubt that people throughout the organization will be ready to change. This, however, is far too simplistic.

In reality, our survey showed that a company's level of readiness to change is determined by a host of variables, including the strength of the corporate culture and how often the organization has engaged in formally managed change efforts before.

At the credit arm of one of the U.S. electronics industry's giants, there was "intellectual awareness" that major change was coming along, but there was "organizational reluctance to change."

Similarly, at a European public telecommunications company, "the greatest risk, of course, was of not doing anything … but it was recognized at the outset that the staff were used to working in a civil service environment and that changing their work habits was not going to be easy."

It is important to get beyond "gut feel" when assessing an organization's readiness for change. It is equally important that the organization's leadership get beyond their own ego in thinking that they have projected a compelling need to change and therefore the company must be ready, willing, and able to change.

A host of assessment surveys have been developed over the years by change management and organizational development professionals. Some are designed to assess the current state of thinking at a company (the organizational culture), while others measure an organization's capacity to face the prospect of change and to accept the actuality of change.

By measuring the climate and giving feedback on the current culture (the organizational as is) to the corporate leadership as they struggle to create a vision of how the company must work in the future (the organizational to be), the survey giver can help the corporate leaders assess the size of the behavioral gap. Then, after measuring attitudes about change, as well as the ability to face and implement change by the individuals throughout the company, the organizational development professional can help the corporate leaders create a plan for facing the resistance and implementing the change.

History teaches us that human beings are extremely flexible and adaptable when it comes to accommodating change in their lives. Nevertheless, individuals and consequently businesses exhibit tolerances or limits to the amount of change they can assimilate over a given period of time.

The futurist Alvin Toffler coined the term "future shock" to describe the threshold beyond which a person or organization can no longer effectively adapt to change. Once this point is reached, healthy coping behaviors are displaced by dysfunctional symptoms such as low morale, miscommunication, reduced productivity, increased anxiety, confusion, high turnover, defensiveness, territoriality, obstructionism, and hostility.

To avoid these costly symptoms of future shock, managers responsible for the implementation of major business decisions need to know what impact change will have on the *targets*—the individuals or groups who will alter their knowledge, skills, attitudes, and behaviors as a result of the change.

Change management is something that cannot be allowed to just happen as part of the BPR effort. You must put together a rigorous change management plan that will run concurrently with your BPR technical changes. Figure 2-6 shows the components of such a plan.

BEST PRACTICE 4: COMMUNICATE EFFECTIVELY TO CREATE BUY-IN. THEN COMMUNICATE MORE.

A number of respondents to our survey said of the BPR effort "you can't communicate enough" to employees throughout the entire effort. This is especially true at the outset.

A good Change Management Plan should include:

Assessment of Change Management Environment
     – Cultural climate
     – Barriers to change

Training
     – Identify skills, gaps, and training needs
     – Develop training materials and workshops/classes
     – Schedule and conduct training workshops/classes

Communications
     – Identification of audiences
     – Translation of vision, plans, and activities into messages
     – Develop communications approach
     – Selection of messages and media/vehicles
     – Communicate developments, changes, and status

Development of objectives and milestones

Figure 2-6. *Components of a change management plan.*

Communication is the most important tool in obtaining buy-in from employees at every level of the company for the changes that will be necessary to reengineer processes.

Motorola has had well-publicized success with both TQM and BPR. For years, the company has held "town hall" meetings to review the state of business with employees. Business unit managers have also held more informal "rap sessions" with employees. The Six Sigma effort (Motorola's quality target) has always been discussed, with much of the focus on the competitive issues and why the goals of quality had to be achieved.

An organization that contemplates change but does not communicate those intentions to its employees is dooming itself, if not to outright failure then at least to a change process that will be more difficult than necessary. As secrets harbored within a family can tear the family apart, secrets in a business organization create anger, tension, and resentment, all of which result in poor business results.

There are two main purposes to an ongoing communication program throughout the BPR effort:

1. To provide communication on a regular basis to people outside the implementation team—on other change teams and throughout the organization—about the changes that will be taking place as a result of BPR.

2. To provide background support for change management activities.

The alternatives to effective communication from the project team are prospects of job loss, problems, and other fears produced by the rumor mill. These stories are always much worse than the actual situation. Both good news and bad news need to be communicated.

At Chevron Chemical Company, whose BPR effort will be highlighted in Chap. 3, there were initial rumors and misinformation. But the company assigned a full-time person just to respond to questions sent via E-mail, making for interactive rather than merely passive communication. Meetings were set up with managers and employees. Weekly updates were sent to each employee via E-mail. A monthly newsletter was sent each employee. And demos of new processes and software were provided.

BPR very often means fewer people to do the same amount of work. There needs to be clear, early communication about policies for what will happen to people who are "reengineered out of a job."

There are also a host of peripheral parties, such as process suppliers and customers, who have a stake in the reengineering activities. A plan must be established to identify all interested parties and ensure that a two-way flow of information takes place.

Effective communication can be used to build support for changes by erecting a framework of honesty and trust on which to build a new business culture.

At Aetna, "it took at least two years for people to believe we had to change. Management felt it couldn't 'tell the children' how bad things were." But realizing that was not effective; Aetna created a position for a PR/communications professional who does nothing but work on BPR.

WHAT'S IN IT FOR ME?

All employees, regardless of rank or salary, and regardless of how they perceive their own security in the organization, want first and foremost to know one thing about any changes that are contemplated: *How will it affect me?*

This question must be answered—immediately, definitively, and by someone who has the authority to provide assurances. Ideally, this person should be the president or CEO, but in large companies it may be a division head or even a plant-level leader. However, he or she must provide assurance from the very top that a job is secure or that the contemplated work or position change is definite.

All employees need to be informed about how the planned change will affect their work group and the company as a whole.

This must be done early in the planning process to avoid rumors, lack of trust, and the perception that the company is engaging in secret activities.

In addition to telling employees about the why and how of corporate change, management must provide regular feedback to employees about the process of the planned changes. One of our respondents, from a major U.S. corporation that does most of its work on government contracts, says his company had to, "make sure top management *understood* rather than assumed that they had buy-in and were communicating. One missing officer at a meeting would result in no consensus." This was visible to all employees, who felt executive support was faltering.

Our survey participants used a number of different communication vehicles to get the message out at the outset of the BPR effort. Fully 70 percent created a company newsletter designed to inform people specifically about the BPR efforts. Many held kick-off meetings with managers or question-and-answer sessions with employees. Forty percent sent individual communications to employees, and 20 percent sent such communications to customers. The full array of communication techniques is shown in Fig. 2-7.

COMMUNICATION FROM THE TOP

The change leader must communicate the compelling need for the change being contemplated, as well as the vision he or she has set out for the company's future state. When a leader shows the organization the urgency of working differently, even when the change itself involves pain, change is more readily accepted.

IBM's CEO, Louis Gerstner, believes that "selling" the new IBM culture to the worldwide enterprise is a fundamental aspect of his job. Since becoming CEO in 1993, Gerstner has traveled extensively, talking to IBM employees and customers.

At first, he announced plans to cut costs by $8 billion through streamlining, outlining the critical need for IBM to become more productive. Then he made the rounds to present his strategies for

What specific communications or events were planned at the outset of the BPR program?

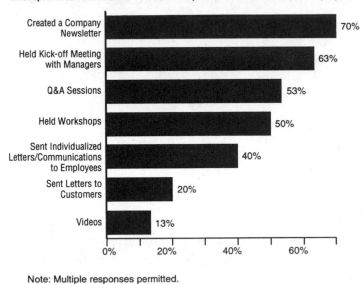

Note: Multiple responses permitted.

Figure 2-7. *Communication techniques.*

responding to the technological shift away from IBM's traditional mainframe and midrange computer business, in order to maintain the company's revenue stream. The third stage of his selling job has been personally championing culture changes he feels are critical to strategic success.

In one typical week, Gerstner's schedule includes trips to Atlanta, Orlando, and Helsinki, Finland, to explain and promote the new ways of doing business.

Gerstner believes changes must be sold by the chief executive face to face. "It's not something you do by writing memos," he says. "You've got to appeal to people's emotions. They've got to buy in with their hearts and their bellies, not just their minds."

When the Orange County, Florida, Corrections Department embarked on major administrative and procedural changes, the director, Tom Allison, met in small groups with every one of the department's 5000 employees. "It took me about six months, doing it in two-hour increments," he says. "But I wanted all of them to hear from me personally what it was that I wanted us to do and how I wanted us to be."

Face-to-face communication, even if not one-on-one, is important because of the opportunity for interaction. This give and take, called *active communication,* is far more apt to achieve buy-in than *passive communications,* such as memos. Too often, passive communication is used to inform employees of change, and it becomes just another missive from the top.

The chance to ask, "What does this mean for me, my group, my friends?" is not necessarily any more reassuring, nor is the message necessarily any easier to take. But the chance to watch the leaders, to judge their body language, and to see their level of personal discomfort over the proposed changes, gives employees a feeling that, "We're all in this together."

Figure 2-8 provides a quick reference for effective communication during a change effort.

Effective Communication

It is impossible to use too much communication.

Simplify your message, no matter how complex the issue. Keep follow-up as simple and understandable as your initial message.

Anticipate the issues, and communicate your position early.

Don't underestimate the technical requirements of a communications project. For especially complicated projects, a full-time communications manager may be required.

Involve top management in delivering your message.

Honesty is the best policy. Tell the truth.

Identify and know your audiences. Select the right message and media for each.

Figure 2-8. *Quick reference for effective communication during change.*

United States Postal Service: Creating Change in a
Mega-Organization

The U.S. Postal Service, created out of the old Department of the
Post Office through the 1970s-era Postal Reorganization Act,
employed 770,000 people in 1994, working in 40,000 post offices
and support centers across the country. This is the second largest
federal organization; only the military, with more than 1 million uni-
formed personnel and hundreds of thousands of civilians, is larger.

Only seven private companies on the Fortune 500 generate
more annual revenue than the USPS's $47 billion. Much like these
corporate giants, the USPS has faced dramatic changes from its
inception to the early 1990s, and remains in a race to remake itself
and remain competitive in the information age.

The USPS mission, according to the Postal Reorganization
Act, is to provide all Americans with quality mail service, including
universal service at uniform prices for at least one class of mail, and
to provide employees with wages, benefits, and working conditions
comparable to the private sector; all the while becoming econom-
ically self-sufficient. These are very ambitious goals, which as of
the end of 1994 had yet to be fully realized.

Clearly, reengineering processes to make them more effective
and efficient, along with increasing the use of automated sorting
equipment, are solutions to improve the overall Postal Service performance. Both of these changes, however, will have a profound
impact on the mostly unionized workforce and on the civil service
model of organizational culture.

Almost 90 percent of postal employees are represented by one
of the four primary unions, which from the end of World War II
until the 1980s were a significant force in an industry characterized
by little competition, few communications substitutes (primarily
the telephone), and captive customers.

But the structure of the industry has changed considerably in
recent years. Facsimiles and electronic mail provide communication substitutes for data over telephone lines. The cost of telephone communications has dropped dramatically since deregula-

tion. New entries into the market to deliver hard copy documents—Federal Express (which changed its name in 1994 to FedEx because that's what customers called it anyway) and a host of others—have chipped away at the more lucrative end of the mail delivery spectrum for overnight mail. City couriers also eat into the cost-effective delivery routes in major cities.

In the nimble, fast paced, ever changing communications industry, the Postal Service is a bureaucracy with 80 percent of its costs tied up in labor costs, and many outdated procedures, some of which were placed there by the very government that is now demanding a "profitable performance." The Postal Service does not even deliver the federal government's overnight mail, because it is prohibited from offering volume discounts and therefore is not price-competitive; FedEx has the contract.

In addition to contending with labor costs and the government, the Postal Service must struggle against its own corporate culture to successfully change. An analysis of the Postal Service's corporate culture in the late 1980s by Duke University identified a culture that was more conservative than innovative; more results-oriented than process-oriented; more task-driven than people-oriented; and more structured than relaxed. It was further characterized as autocratic, internally focused, functionally driven, and not strategic in outlook. Marketing was perceived as weak.

In 1992, Marvin Runyon entered the picture as the 70th Postmaster General of the United States. Within weeks of joining the agency, he announced plans to eliminate 30,000 jobs of management employees who "don't touch the mail." He offered buyouts and in fact 47,838 employees accepted. The problem was that only about 16,000 were managers; the rest were experienced craft employees, many of whose jobs have had to be filled by inexperienced new hires to keep operations running.

By late 1994, two years into Runyon's tenure, the biggest problems have yet to be resolved. The Postal Service continues to lose money—$1.7 billion in fiscal 1993 and an estimated $900 million in fiscal 1994. There were more postal workers in late 1994 than when Runyon took over in 1992. And delivery problems in New York, Chicago, and Washington, D.C. have been widely publicized.

Where Runyon may have been more successful is in his reorganization of the executive levels of the Postal Service to create a new emphasis on customers. Even changing the nomenclature from bureaucratic to corporate—such as changing the titles from Associate Postmaster General to vice president—is a subtle but important piece of the overall effort to change the corporate culture.

Runyon has started to change the language of the Postal Service by asking executives to look at competition and to seek improvements not just in cost but in service. He is forcing the Postal Service to begin measuring its success in terms of customer service. He continues to emphasize aggressive marketing and communications; he has hired the former White House spokesman, Larry Speakes, as Vice President of Communications, and the former CEO of CitiBank of Illinois, Loren Smith, as Vice President of Marketing.

In its 1993 Annual Report, the USPS enumerated its five Guiding Principles, which speak in the language of business organizations doing business in the highly competitive, customer-focused and team-oriented ways of the 21st century. These principles are shown in Fig. 2-9.

What role does BPR play in this environment? How can managers and employees throughout the organization hope to achieve the excellence that is one of the five guiding principles for the Postal Service when the individual with the most power, the Postmaster General, has difficulty effecting change?

Business Process Reengineering at the USPS is currently organic. Several reengineering projects are in process in a variety of departments and at a number of sites throughout the Postal Service. Some are further along than others. Many of those efforts are coordinated through the Information Systems Organization, which has developed its own framework for undertaking BPR projects, the basics of which are to:

Create a charter.

Define the as-is of the process to be reengineered.

Develop a vision for the new process.

Develop alternatives for achieving the vision.

Develop specifications for the selected alternative.

UNITED STATES
POSTAL SERVICE™

Annual Report of the Postmaster General
Fiscal Year 1993

Guiding Principles
The Postal Service is committed to:

People

Diversity is valued; everyone must be treated with dignity and respect. Training and information must be provided to employees. Preparation strengthens teamwork and participation in decision making, which are essential to customer and job satisfaction.

Customer

We will achieve the highest possible levels of satisfaction with every service encounter. Customer satisfaction is essential to the health and growth of our business.

Excellence

We stand for continuous improvement, positive change, and making breakthroughs in what we do and how we work. Each of us will bring our finest efforts to bear on each task and each endeavor, all the while looking for better, easier, faster, and simpler ways to serve our customers, achieve our goals and improve our performance.

Integrity

We will be worthy of the trust given us by the American people. We will act with integrity in every encounter and relationship with postal customers, business partners, and each other.

Community Responsibility

We will build upon our legacy of more than 200 years of service to the nation by meeting the changing needs of the communities we serve into the next century.

Figure 2-9. *USPS Guiding Principles.*

Perform change management.

Develop an implementation plan.

Implement the new process.

Measure and monitor the process.

A project is typically chartered at the vice presidential level, or one level below, and the person who charters the project becomes the executive sponsor. A management review board is chosen, followed by a project core team to actually reengineer the process. On the management review board are executives from the functions that are stakeholders in the process to be reengineered.

A Postal Service employee chairs the project core team, and consultants, if used, work closely with the project core team. The project core team reports all its results, including information gathering and recommendations for reengineering, to the management review board.

As of the end of 1994, the Postal Service had more than a dozen BPR initiatives underway. Each looks for "quick wins," which can yield cost savings, improve the work processes, and increase morale over the short term. These successes allow BPR teams to push ahead with long-term opportunities that will yield even more dramatic improvements in the business process.

Creating change at the Postal Service is a multipronged challenge. To some degree, change has occured in that:

A focus on customers is beginning to emerge and become the focus for reengineering efforts.

Stronger, more cohesive teams are formed that stay together throughout the reengineering effort. In early BPR undertakings, teams often drifted apart and the process change was implemented by a few people who "stuck with it." Often these changes didn't last.

But to drive organizational change into the fiber of an organization the size of the Postal Service, a succession of successful BPR projects will need to be accomplished. Only when enough processes are made more effective and efficient can meaningful downsizing and reduction in head count take place.

Let's look more closely at one postal service BPR effort.

Express Mail competes head to head with overnight carriers such as FedEx, DHL, and United Parcel's Overnight Letter. In 1993, 53 million pieces of Express Mail were delivered, contributing $627 million in revenue. (Source: USPS 1993 Annual Report)

Reengineering of Express Mail began in April 1993 and focused primarily on finance and customer support. A task force discovered a number of problems with Express Mail processes, including duplication of effort, confusion about job responsibilities, and activities that added work with no real benefit.

The Express Mail reengineering project was chartered in the fall of 1993, with final recommendations presented to the management review board overseeing the effort on December 22, 1993. The initial project focused on "administrative support" processes, called Expedited Services, not on processing and distribution of the physical pieces of Express Mail.

Although these are not the processes most people associate with the Postal Service—receiving, sorting, transporting, and delivering Express Mail packages—administrative support processes directly touch customers and have a significant impact on customer satisfaction.

The project's goals were to:

Identify short- and long-term improvements to Expedited Services.

Streamline Exedited Services to improve utilization of resources.

Eliminate unnecessary work associated with Expedited Services.

Enhance current customer service levels.

Develop a functional vision for the Expedited Service Office, of which Express Mail is only one product.

Create a "roadmap" to achieve the vision.

Team members considered to be the "best and brightest" were assembled and expected to devote 75 to 100 percent of their time to the team's work. Union members, though not actually team members, were encouraged to provide input into the process. The team was given physical space in which to work, training as needed, and a facilitator from an outside consulting firm.

Five major processes were identified:

Customer inquiries.

Corporate account setup and maintenance.

Performance analysis and improvement.

Label handling, data entry, and verification.

Express Mail technical and design support.

Within these five processes, 32 subprocesses were identified and recommendations were made for each. Many subprocess changes were "quick hits"—70 percent of the recommendations were implemented within three months, and they account for 50 percent of the cost savings recognized in the project. Fig. 2-10 shows the processes reviewed by the reengineering team.

There are a number of change management lessons learned from the Express Mail reengineering project.

First, the details of changes to take place in Express Mail were initially not communicated widely, for fear that too much communication might cause unnecessary confusion and resistance to the change effort. However, as the Postal Service becomes more confident about the benefits gained from BPR efforts, the organization is likely to communicate more and earlier in future projects.

Second, referring to the Burke-Litwin model, you can see the different levels of change attempted at the Postal Service. Runyon's downsizing and reorganization are transformational changes, requiring strong leadership and direction. In contrast, the type of changes recommended by the Express Mail reengineering project are transactional.

While we generally advocate transformational change, in some instances the corporate culture makes this especially difficult. Rather than have no change, transactional changes that produce significant results but that do not require the same type of leadership as transformational changes can be implemented. Individuals at various levels and at different locations in the organization can make successful changes of this sort, demonstrate success, and help drive the organization toward future transformational changes.

Choosing to focus on finance and customer service support as well was important to a successful reengineering effort in Express Mail. Many more employees are involved in processing and distribution, and therefore it is much harder to successfully implement change. This is true not only in the Postal Service, but in any large, unionized organization.

Because the most cost-effective gains could be secured on the business side of Express Mail, and because implementation would be somewhat easier, this was the logical place to focus.

UNITED STATES POSTAL SERVICE.

Processes Reviewed

Customer Inquiries
Process Delivery Inquiries
Process Product Inquiries
Process Financial Inquiries (EMCA)

Corporate Account Setup and Maintenance
Set up New Account
Cancel Account
Deposit Funds into EMCA
Monitor Financial Status of EMCA
Process Credits to EMCA
Obtain New Accounts
Monitor Account Status–Sales
Send Monthly EMCA Statement

Performance Analysis and Improvement
Perform Destination/Originating Failure Performance Analysis
Analyze CTT Daily Transmission Report
Expand On-Demand Pickups

Label Handling, Data Entry, and Verification
Process Refund for Prepaid Customer
Collect on Short Paid Items
Perform EMCA/CD/Federal Agency Label Verification
Analyze EMRS Rejected Labels
Process Express Mail Labels
Audit Daily IRT Transactions
Audit IRT Transmission Report

Express Mail Technical and Design Support
Distribute Supplies
Facilitate Drop Shipments
Facilitate Mail Reshipments
Facilitate Custom Shipments
Coordinate On-Demand Pickup
Maintain Network Information
Maintain CTT/EMRS Database Tables
Provide Training
Support CTT Hardware
Order CTT Hardware
Process IRT Close-Out

Figure 2-10. *USPS processes reviewed.*

PROSPECTIVE BEST PRACTICE: INSTILL IN THE ORGANIZATION A "READINESS AND COMMITMENT" TO SUSTAINED CHANGE.

If they do nothing else, Marvin Runyon's efforts at the Postal Service are beginning to create an organization that understands change and that is ready to meet change, even if it is not an organization that exactly welcomes change.

The single most important factor necessary to increase an organization's speed of change and ability to continuously change is the degree to which people are resilient. *Resilience* is the ability to absorb high levels of disruptive change while displaying minimal dysfunctional behavior.

It is no longer sufficient to merely adapt to new demands, cope with the stress of uncertainty, or adjust to disruptions in the workplace. Resilience is the force that allows people to go beyond survival and to actually prosper in an environment that is becoming increasingly complex.

Although resilient people face no less of a challenge than others when they engage change, more often than not they:

Regain their equilibrium faster.

Maintain a higher level of productivity.

Are physically and emotionally healthier.

Achieve more of their objectives than people who experience future shock.

Tend to rebound from the demands of change even stronger than before.

Resilient people are positive, focused, flexible, organized, and proactive.

PROSPECTIVE BEST PRACTICE: STAY ACTIVELY INVOLVED.

"Staying involved" can be regarded as "walking the talk," and it is the behavior that separates the real leaders from the figureheads. Because top executives have so many responsibilities, it would be easier if their job in leading change was over once they had articulated a vision and instilled it in their top managers, who then could disseminate it throughout the organization.

But that is not the case. Successful change management requires their continued actions as champions, role models, and overseers of change. Their active involvement may include chairing the steering committee or participating on it, presiding over ceremonies where employees are rewarded for their adherence to the new behaviors, continuing to communicate in large and small forums, and visibly adopting the new behaviors being asked of everyone in the organization, such as participative management, focusing on processes, and making fact-based decisions.

In Indianapolis, Mayor Stephen Goldsmith came to office with a mandate to improve city services and cut costs. When he told city departments that they would have to bid against private contractors on the basis of cost and quality, street repair employees told him they could win any competitive bid if he would get rid of the layers of patronage "supervisors." Despite the ire of local party officials, Goldsmith dismissed more than half the political appointees.

When AlliedSignal's newly appointed Vice President of Quality and Productivity laid out a five-year plan for CEO Larry Bossidy, Bossidy reminded him of the company's aggressive goals with regard to speed. He told the VP to get the job done in two years, prepare to close his office in three years, and assume an operations job. Said one AlliedSignal executive, "The obvious commitment from the top gave all of us a sense of personal urgency about making the changes."

ORGANIZING FOR SUCCESSFUL BPR

Best Practice 5: Create top-notch teams.

Best Practice 6: Use a structured framework.

Best Practice 7: Use consultants effectively.

Prospective Best Practice: Pay attention to what has worked.

B ecause a BPR effort is not about command and control, organizing for that effort is not about creating a hierarchy. BPR is about making business changes. To do that, you need to work through teams, focus groups, and interaction among individuals to develop and share ideas.

Four groups of individuals are essential and required for BPR:

1. Executive steering committee.
2. Reengineering work teams (sometimes called process teams).
3. Line management.
4. Facilitators and/or consultants (internal or external).

The *executive steering committee* is led by an individual at an appropriate level of authority to actually implement changes, often the CEO. This steering committee is made up of top executives who champion the change effort and who sponsor the change to all staff. It also sets improvement targets.

Committee members assign resources to the reengineering task, strive to remove barriers to progress, and work to assure integration of the BPR effort with other company initiatives. They choose team members for the teams that will do the planning, reengineering, and implementation work as the project moves forward. As a group, members set policy and targets, and make any decisions on moving the effort forward. This is also the group that monitors results throughout the effort.

Reengineering work teams are cross-functional and are pulled together to carry out the analysis and to develop the recommendations for change. These teams will change in membership as BPR moves from planning toward implementation. In the planning stage, a higher-level, broader-based group leads the analysis, which is then presented to the steering committee, which selects the processes for reengineering. Once a process is selected, a group that is more expert in the technical subject matter performs the detailed analysis for reengineering.

After the work teams have evaluated the current situation and given the steering committee recommendations for processes to reengineer, the steering committee develops a process vision for each process chosen for reengineering. The more detail-oriented reengineering work teams then reengineer the process, develop process measurements and targets, and finally develop full implementation plans. Along the way the reengineering teams make presentations to the steering committee, at which points senior executives explicitly "sign off" and recommit to an ongoing effort.

Members of *line management* contribute resources to teams and work to implement short-term improvements. They also provide input to teams as internal customers. They participate selectively in executive workshops, and in the end they are responsible for implementing the reengineered processes.

The reason for these teams and for intense communication between teams is twofold. First, the best ideas come from groups of people, working together in good faith, to explore potential changes and solutions to current problems. In such conditions, people bounce ideas off one another, refine each other's ideas, challenge each other's ideas, and otherwise augment one another's efforts. Second, even if a reengineering work group comes up with a terrific idea, it cannot move forward until there is a certain amount of "buy-in" from above, an explicit acknowledgment that this is an idea worth moving forward with.

Any changes in processes will have a large impact on line managers, who will necessarily feel challenged by and possibly ill at ease with what is going on around them. Successful change requires the participation of those who will be affected. The key is to take the steps necessary to involve them and to communicate constantly as the BPR effort moves along.

One way to involve line executives is to include some on reengineering work teams. For those not intimately involved with teams, "preselling" the new vision is important. Preselling can be accomplished by having key line executives involved selectively in the periodic progress-update meetings with the steering committee.

Facilitators and consultants help deliver results. They can be either outside consultants or internal professionals. They work to coach the teams in group dynamics, and facilitate executive workshops and team meetings. They provide tools, techniques, methodologies, training, and expert-level reality checks on conclusions and direction as the effort progresses.

Figure 3-1 shows the roles and responsibilities of the four main groups.

As you can see by now, BPR is truly a team-based undertaking. In contrast to the old-fashioned way of doing things, BPR calls for planning and implementation to be broken down into tiny tasks and assigned to individual team members to lead or coordinate, thereby moving forward by a momentum formed of teams coming to consensus-based conclusions. Team members and subgroups perform tasks between meetings, but they are usually information-gathering tasks, designed to bring to the next team meeting more information with which the team can take action.

How teams work is therefore critical.

Executive Steering Committee	– Develop overall business vision – Approve reengineering targets and develop process vision – Champion the change effort – Remove barriers – Provide initial guidance on team membership – Monitor results
Reengineering Work Teams	– Work with Executive Steering Committee to finalize process vision – Perform "as-is" analysis – Develop "to-be" model – Develop and redesign metrics – Create implementation plan
Line Management	– Provide appropriate team members – Contribute resources – Implement short-term improvements – Provide input to teams as internal customers
Facilitators/Consultants	– Provide reengineering framework to be used by teams – Provide Just-in-Time training on tools and techniques – Encourage team members to think "out-of-the-box" – Play devil's advocate for improvement suggestions – Push to reach stretch goals

Figure 3-1. *Roles and responsibilities of the four main groups.*

Plans are formulated through a series of all-day meetings, sometimes called *workshops*. Each such meeting is structured so that an outcome or conclusion can be facilitated. Between meetings, team members engage in fact finding, analysis, and building business cases in preparation for the next meeting. Training for team members is conducted in a JIT manner, just before a tool or technique will need to be used.

Support by corporate leadership for the team concept and support of the teams as they hit the bumps in the road are key to the success of any BPR effort. Leaders need to build morale and a positive attitude toward change, allocate resources, and support the teams as they are constituted. They also reserve the right to intervene and settle disputes, as well as to set the course as necessary.

The proof of the teaming concept is in its success. One company executive responding to our survey said simply, "The contribution of the team was greater than any possible contribution of the three individuals working independently."

BEST PRACTICE 5: CREATE TOP-NOTCH TEAMS.

There is a story told about Werner von Braun, a pioneer of the American space program. Someone once found him lying on the beach at Cape Canaveral, staring out at the waves.

"What are you thinking about, Dr. von Braun," the visitor asked. "Are you dreaming about an innovation in space flight or how to solve some problem?"

"No," von Braun replied. "I am thinking about something much more important, my team."

Just like von Braun, corporate leadership needs to spend a lot of time thinking about having the right people on the process teams that will undertake BPR. Having the right teams is possibly more important than finding the right solution. There are a couple of reasons for this. First, the right team probably will come to the right solution, and more expeditiously than an individual. Second, the right team will be able to create buy-in throughout an organization for the solution.

Dramatic change requires people with creativity, vision, and openness to innovation. Often they will already be some of the busiest people in a company because everyone recognizes and respects their talents. Yet they are made available to the BPR effort because the company's leadership gives this assignment its highest priority.

Of the 47 companies in our survey who thought they had progressed reasonably far down the BPR path and who believed they had achieved some measure of success, all of them used teams. Nearly three-quarters of the teams were appointed by a steering committee made up of senior executives. Division heads were always on the team. Fifty-four percent had internal company experts on the team, and 47 percent used outside consultants. Only 16 percent had IT staff on the team. Staffing of the initial project teams is shown in Fig. 3-2.

One company's executive told us of its criteria for team membership:

> The project team has approximately ten members who are dedicated to the project. Included are two to four consultants. The team leader was selected because [he] was a champion

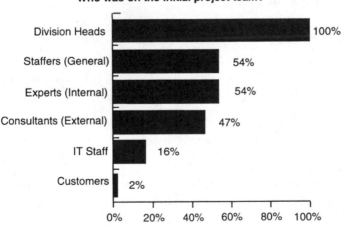

Who was on the initial project team?

Note: Multiple responses permitted.

Figure 3-2. *Staffing the initial project teams.*

for this type of change. Team members represented sponsors of the project. Team members were nominated because of key knowledge that they brought to the team. Senior managers were briefed to get approval for employees in their organization [to participate]. International representation was critical.

Another company, skilled in teaming techniques from previous TQM work, had over 4300 teams working in 1993, many of whose members were volunteers. These "total customer satisfaction" teams were self-organized to focus on areas where they can make improvements, then break up after achieving their goals.

Teams formed for the initial stage of BPR, during which time the work team analyzes potential processes for reengineering, should be as broad-based as is practical, with representatives from operations, finance, corporate strategy, human resources, and information services. Those who are not formally on the team can also contribute to the planning and formation of a BPR effort. Team members can either interview them one on one, or they can participate in focus groups. These interviews can help in:

Describing an as-is process.

Assessing/defining corporate strategy.

Appraising current performance and organizational fitness.

Describing the corporate culture and identifying key change-management issues.

Validating/reality-checking a BPR vision.

Remember that, if you ask for input, you need to respond to suggestions. Everyone who contributes time to this interviewing and idea gathering must receive a copy of the final plan, and many if not all should be invited to briefings before the general roll-out briefings.

A Team Is More Than a Group

Our survey showed a decided bias among companies for choosing team members who are experts (97 percent chosen for their expertise) and stakeholders (83 percent chosen because they were stakeholders). Other reasons for choosing team members are that they have positions of authority, they are creative thinkers, they have good project management skills, or they are good "team players." Figure 3-3 shows personal characteristics that led team members to be chosen.

After a BPR effort, survey respondents were asked to assess the qualities and characteristics of teams and team members that led to team success. Forty-three percent said the fact that team members were the "best of breed" contributed significantly to team success. Forty percent said success was enhanced because communication was strong among team members. But the most critical factor in achieving success was the team's ability to maintain a focus on the BPR vision and objectives; 62 percent cited this as the key factor in achieving a successful team effort.

But regardless of who is asked to participate on a team, it is important to remember that a team is more than merely a group.

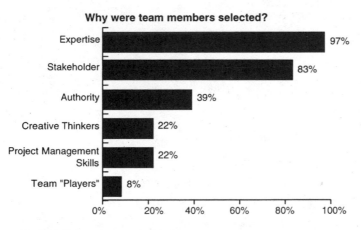

Note: Multiple responses permitted.

Figure 3-3. *Personal characteristics of team members*

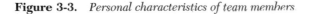

In very few instances could an all-star professional sports team beat the team that wins the league championship in a series of games. While the all-star team is filled with "best of breed" players, the championship team has honed its team skills over an entire season and a series of playoff games to win the championship.

A reengineering work team should be more a championship team in the making than an all-star team. The talent needs to be there to reengineer the process, but the team skills need to be developed to work together consistently over the long haul. It takes work to transform a group of colleagues into a true team. Figure 3-4 shows some characteristics of effective and ineffective teams.

Getting the right players in the first place can help pave the way for a championship team. At Aetna, team formation "became part of (the) internally developed methodology." The process of

---

What Makes an Effective Team?

In Effective Teams:

Communication goes two ways.

Members openly and accurately express all their ideas.

Team members share participation and leadership.

Decision-making procedures are appropriate for the situation—teams discuss issues and try to reach consensus on them.

Constructive controversy and conflict enhance the quality of decisions the team makes.

Members evaluate the effectiveness of the team and decide how to improve its work.

In Ineffective Teams:

Communication is one-way.

Members express their ideas, but keep their feelings to themselves.

Member participation is unequal: members who hold positions of authority tend to dominate.

Decisions are always made by members who possess the most authority—there is minimal team discussion.

Controversy and conflict are ignored or avoided, and the quality of decision making suffers.

The highest-ranking member of the team, or management itself, decides how to improve the team's effectiveness.

Figure 3-4. *Characteristics of effective and ineffective teams.*

team formation included talking to all sponsors of the BPR effort and getting the best and brightest on the team, as well as representatives from each SBU. In addition, teams are to include people who have been around long enough to have seen changes and who can be defenders of what is good and worth maintaining. The company also seeks analytical people and visionary people.

As soon as people know they are going to be placed on a team, they are given background reading. Then there is a kick-off meeting, at which discussions are held about the nature of reengineering and the entire framework. People from other teams are brought in to share their experiences; there is also training in benchmarking, home-grown tools, templates, and other areas. Throughout the effort, team members are coached and educated, "learning while doing."

TIME

How much time does it take to make a BPR breakthrough? For simple business processes confined to a few functions, it could take as little as three or four months. To reengineer more complex processes that cross more functional boundaries—as most core business processes do—it may take well over a year to complete the implementation.

Simple or complex, a breakthrough is a major change initiative that requires focused attention. Members of the initial design team should count on a full-time commitment during the process-mapping and creative planning phases, which usually takes three to four months.

Specific process reengineering teams may be extremely active for six months or more during the implementation. The steering committee will be very active early on in the effort, with less frequent and less intensive reviews of the undertaking once it is into the implementation. Figure 3-5 plots graphically how the intensity of the many teams' workloads varies over time during a BPR effort.

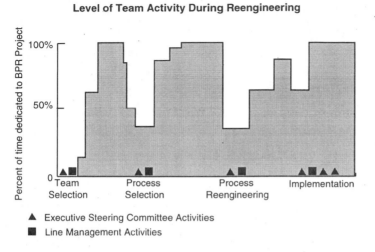

Figure 3-5. *Variations in workload.*

In our survey, 42 percent of respondents said team members spent between three-quarters and all of their time on BPR; another 22 percent spent between half and three-quarters of their time on BPR. A full breakdown of time spent on the BPR effort is shown in Fig. 3-6.

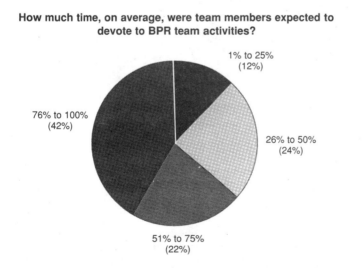

Figure 3-6. *Time spent on BPR.*

Interestingly, about one-third of the respondents felt that not enough time was spent by their team members. Another third of the respondents cited "enough time spent" as a critical factor in team success. However, the amount of time actually spent did not always correlate with overall project satisfaction. Dataquest has suggested that the question, "How much time is enough?"—in terms both of percent of time spent by team members and of chronological time to complete a project—is worthy of more study. We agree.

TEAM DEVELOPMENT STAGES

Teams go through various developmental stages over time. We define four major developmental milestones:

1. Team formation
2. Conflict
3. Home base
4. Synergy

Others call these phases form, storm, norm, and conform.

TEAM FORMATION. As teams form, members generally feel excited and optimistic about working together. They often act tentatively, and, whether or not they show it, they are somewhat anxious about the work ahead. They are primarily concerned with the answers to four questions:

1. What is the team's purpose?
2. What methods and procedures will we use?
3. What will be acceptable team behavior?
4. How will we be judged?

CONFLICT. During this stage, team members are likely to argue about what actions the team should take and what methods should be used. In fact, they'll often resist collaborating.

In this stage, some team members begin to express concerns about the team's chances for success. There is resistance to trying different approaches, and team disunity may prevail.

At some point, team members begin at least to understand one another, even if they are not completely on track together moving toward their goals. During this stage, members are focused on finding answers to three questions:

1. How should conflict around the team's purpose and methods be resolved?

2. How should the team deal with such problems?

3. How will roles be decided?

HOME BASE. During this stage, team members decide to accept team rules and roles, as well as fellow team members' strengths and idiosyncrasies. By this time, the team is doing its job. Team members feel a sense of personal accomplishment, belonging, and mutual trust. They also feel free to express ideas and to give and receive constructive criticism.

Most productive teams spend much of their time in this stage.

SYNERGY. By this final stage, the team is an effective, cohesive unit that gets a lot of work done. While moving through the synergy stage, team members feel creative, trust one another, and take pride in the team's accomplishments. Members' activities and abilities are effectively coordinated. Teams in this stage make decisions by consensus and communicate well with people both inside and outside the team.

As teams become adept at working together—and maybe even begin to feel like a "family"—they need to watch out that they don't fall into the trap of "groupthink." Team members need to make sure that they engage in fact-based analysis and continue to challenge one another. If the team has concentrated on arguing about the facts from the beginning and has left personalities aside, this will be easier.

CHARTERING THE TEAM

Teams will be most effective if they have marching orders, sometimes stated formally in a team charter. In our survey, we found that only 21 respondents—fewer than 50 percent—had any kind of charter and that, when they did, it was usually broad, to "meet the overall goals of the BPR effort."

We believe a team charter should be more helpful than that, although it should not be all-consuming. At its most straightforward, a charter should:

> Delineate the task.
>
> State management's expectations.

The task should be narrower than the company's strategy statement, and the expectations should tie goals and metrics to the task. For instance, the charter may ask reengineering work teams to:

> Reengineer a core process that will help the company increase our customer satisfaction ratings by 25 percent over the next 12–18 months by means of a reduction in lead times and better aftermarket service.

Remember, however, the charter can always change if there is a need for a midcourse correction.

Along with properly chartering teams, senior executives can do a lot to help teams succeed, from the initial team selection through the entire BPR effort. Figure 3-7 gives some practical pointers.

PROJECT MANAGEMENT

Success in a BPR effort can be enhanced through thoughtful and rigorous project management. Project management creates overt links between the technical changes that will be made to work processes and the behavioral changes that will need to be made to

Helping BPR Teams Succeed:
Practical Pointers

For continuity, create core teams that will participate throughout the BPR process. You can get special skills you need on a short-term basis by forming ad hoc project teams to complete discrete tasks. And while the core of participants should remain the same, the project structure can change as needed.

Choose team members with experience in strategic visioning, change management, and team improvement initiatives. People with varied backgrounds, even those without direct experience in the core process to be changed, are important because they can generate new insights and challenge the status quo more rigorously. They should be the "best and brightest" in your organization.

Balance creative and change management skills on a team.

Make sure that internal or external BPR support team staff think through "out-of-the-box" ideas in advance of creative sessions. This will "seed" brainstorming for even more creative solutions.

Give your teams the power to change anything as they develop objectives and designs.

Give your teams the training they need to conduct BPR activities effectively. Examples include training in how to use simulation tools, process-mapping, and group decision techniques.

Build commitment to the redesign by involving people with credibility and influence in your company—at all levels of authority and across all units.

Leverage experience that already exists in the organization, especially in departments with background in training and development.

Conduct team-building workshops/events if needed to bring team members together and overcome barriers to cooperation.

Be prepared to add outside suppliers and even customers to teams when their input adds value to the BPR project.

Figure 3-7. *Pointers on making the BPR effort succeed.*

make the technical changes a reality. It also creates a vehicle for project communication.

Project management will work only as a participative exercise. It is not a command and control activity. There should be project management at the overall BPR effort level—often done by a person called a *program manager.* And there should be project management for each BPR work team.

Some companies like to use outside consultants for their project management, while other companies feel it is important that project management be "owned" by insiders, with any outsiders being used to train project managers, if necessary, and possibly to facilitate teams. Some companies have regular meetings of their project managers so that they can all pass around useful information, both on the progress their teams are making and on successful techniques they have used to break through any problems.

We like to think of the right way to approach project management for a BPR effort as Goal-Directed Project Management, a technique that we have used for our own consulting staff and that we have trained client project managers in.

All projects in a business create change, and BPR creates massive change. Only by having defined goals and objectives linked to a clear business need can change be successfully implemented. The project manager becomes the fulcrum by which executives leverage the teams' efforts.

In addition to planning that is focused on meeting overall goals and objectives set out for the effort, a successful project manager needs to engage in milestone planning and activity planning. This is true both for the project as a whole and for each subproject.

Goals and objectives, the first tier of project management, define what, not how. They are the ends, not the means of getting there. These goals and objectives must provide descriptions that do not change throughout the life of the effort, and there should only be a few of them. These goals and objectives must be:

Within the team's control.

Measurable.

Results-oriented.

Meaningful.

The second tier elements, *Milestones,* are derived from the project's goals and objectives. They represent intermediate states that must be achieved to meet the project goal. They should be expressed as conditions to be met or satisfied, not as actions to be taken; again, they are what, not how. They need to be agreed to up front and firmly scheduled.

Effective milestones have a natural flow and represent important decision points. Again, they must be controllable and limited in number. And they should occur at useful intervals.

To effectively plan for milestones, you should involve those who are likely to be responsible for completion, and avoid focusing on activities necessary to reach the milestones. Milestones should be neutral with respect to solutions; in other words, the way a milestone is worded should not imply a particular solution, which often hinges on particular activities. Finally, all the project's objectives should be addressed within the map of milestones.

Activity planning is the third tier of project management. In activity planning it is important again to involve those who will be committed to the work in planning the activities. It is also important that the activities planned be controllable. And it is useful to adopt a "rolling wave" approach, where activities group together in waves and create their own momentum.

BEST PRACTICE 6: USE A STRUCTURED FRAMEWORK.

Two-thirds of the companies we surveyed used a structured framework or methodology in the Business Process Reengineering effort; of those, 60 percent used a methodology designed by an outside consultant, 20 percent a methodology developed in house, and 20 percent a combination of consultant-designed and in-house methodology.

"Methodology" seems too mechanical to some people—including us—and can give people the idea that there is a cookie-cutter way of conducting a reengineering effort. A good methodology, however, will provide you with a structured framework. It will facilitate understanding and communication by breaking the effort up into recognizable pieces and by having a common language in place with which to discuss reengineering.

We prefer to use the word "framework" and will refer to frameworks, not to methodologies.

The advantages of an in-house framework are that it comes from a familiar cultural base and often presents ideas in a way employees are already familiar with. The disadvantage, of course, is that too much familiarity could allow people to slide through a reengineering effort without ever trying to shatter the paradigms within which they currently operate.

The advantages of using an outside framework is that, in the best of cases, it is based on breadth of experience with many different companies. The disadvantage is that, if it is not presented well or constantly revised to take into account the rapid changes in business, it can take on the feeling of a cookie-cutter approach.

We have often found that the best results occur when a company brings in an outside consultant, learns and understands the consultant's approach—while the consultant is learning and understanding the company's specific circumstances, language, and culture—and then incorporates some of its own language and culture into the broad parameters of the consultant's approach.

AlliedSignal has accomplished great strides in reengineering a huge company under the banner of Total Quality. It did so because of the ongoing effort that had gathered a lot of internal support up and down the company, and under which people had developed common language, common culture, and common goals and visions for the company. Reengineering built on that.

At Aetna, the company's internal framework includes methods for:

Project Selection.

Project planning, which includes requirements for:

Defining a mission.

Defining critical success factors.

Internal and external scans.

Defining gaps today, and predicting gaps to be filled in the future.

Objectives (what will be delivered? to whom? why?)

Steps to be taken.

Team formation.

Project management (done by a reengineering team).

One company in the survey told of creating its internal framework after talking to a number of customers and suppliers about what has worked for them, as well as calling in a number of consultants to "make their pitch and talk about their approach."

Whether you develop your own framework for action, use one supplied by an outside consultant, or create a hybrid, Fig. 3-8 lists the important elements of any good framework.

A Successful BPR Framework:

Incorporates change management.

Provides for organizational communication.

Allows for radical change.

Prescribes clearly defined goals/targets.

Provides a variety of "tools" to be used throughout processes as necessary.

Plans for customer/supplier input.

Integrates IT.

Is flexible enough to be tailored to the organization's needs.

Figure 3-8. *Elements of a good framework.*

BEST PRACTICE 7: USE CONSULTANTS EFFECTIVELY.

Support from a variety of internal or external consultants is sometimes required to apply the variety of disciplines needed to plan and oversee a BPR effort. While team members are often chosen to supply some of this expertise, there also needs to be people who stand outside the team structure, who can act as:

Coaches, who can offer leadership, encouragement, and an experience-based assessment of what it will take to make BPR happen.

Facilitators, who use proven tools and techniques to keep the change process running smoothly.

Visionaries, who can focus more freely on the future because they are experienced innovators—and because they have no stake in the past.

Experts, who have the knowledge and skills to conduct BPR.

Project managers, who have the time, the tools, and the experience to coordinate diverse, broad sets of corporatewide activities.

Trainers, who can quickly instruct a company's staff in the day-to-day skills needed for BPR, such as process mapping and simulation of the as-is state, using statistical improvement tools in process design and measurement in the new processes, and in creating the vision necessary to start the BPR effort and keep it focused.

Some companies have these resources in house. Others choose to develop them. Still others prefer to bring in assistance from the outside.

In our 1994 survey, 76 percent of respondents used external consultants, in various roles, though not always throughout all project stages. Sixty-six percent used consultants to help with management strategy, 63 percent in project management, 34 percent in IT implementation, and 31 percent in change management areas and in training. Figure 3-9 shows the frequency with which outside consultants were used and the roles they played.

Consultants were least often involved in the implementation of BPR; many respondents felt it was important that their company "own the process" of BPR. As one survey participant said, "The idea that you can bring them in and 'let 'er rip' doesn't work. [Our] users should own the project."

When consultants were used in implementation, it was most often in the IT area.

While 30 percent of our respondents considered consultants critical and 46 percent considered them extremely critical to the BPR effort, only 33 percent considered the consultants they used to have been extremely effective, while 39 percent considered the consultants they used to have been generally effective, as shown in Fig. 3-10. Use of consultants was not a guarantee of an effort perceived as successful, and some of those companies that regretted not using consultants still felt their BPR efforts were successful.

Consultants were heavily used in providing training for teams. Eighty-three percent of respondents provided general BPR train-

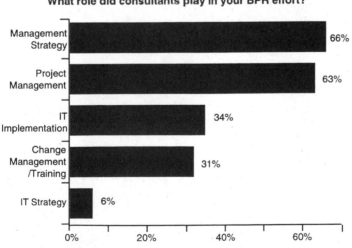

Figure 3-9. *How often, and how, consultants were used.*

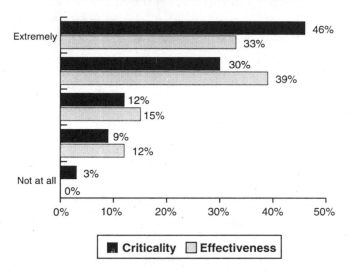

How critical was the role consultants played in the reengineering effort, and how effective were they?

Extremely — 46% / 33%
— 30% / 39%
— 12% / 15%
— 9% / 12%
Not at all — 3% / 0%

0% 10% 20% 30% 40% 50%

■ Criticality □ Effectiveness

Figure 3-10. *Results of using consultants in the BPR effort.*

ing for their teams, while 38 percent provided training in project management and 32 percent in benchmarking.

Training was also provided in the group dynamics area, with creative problem-solving training provided to 27 percent, communication skills to 27 percent, and negotiating skills to 24 percent.

Data analysis, data collection, prioritization, simulations, competitive analysis, and financial analysis were also provided to anywhere from 10 percent to 25 percent of respondents' teams.

Figure 3-11 provides you with some hints for choosing the right consultant to help you.

PROSPECTIVE BEST PRACTICE: PAY ATTENTION TO WHAT HAS WORKED.

The best BPR undertakings build on previous efforts and programs, especially those with a Total Quality focus. Other corporate efforts that have made people more receptive to change and more

Choosing a Consultant

To be sure you are getting a
well-qualified consulting firm that is right for you,
look for these strengths:

A knowledge of and experience with the BPR process in your industry.

An understanding of your business and marketplace.

A track record for real innovation (not just for recycling or relabeling old ideas).

A flexible approach that ensures tailoring BPR to your unique situation.

The diversity and depth of staff experience in all the cross-functional disciplines involved in BPR. The wider the variety of resources available to you, the better.

The willingness/ability to commit their senior consultants to work with your top managers on this high-level project.

A wide variety of effective BPR tools and techniques to apply, and experience in using them.

Experience in team-building and BPR project management.

International experience, if you have—or may want—an international market operation.

Sensitivity to the effects dramatic change will have on your managers and workforce.

Expertise in both BPR and information systems reengineering.

Good "chemistry" between your people and the consultants.

Figure 3-11. *How to choose the right consultant.*

willing to experiment and challenge the definition of their work will make them more willing and able to engage BPR both philosophically and tactically. Companies that have stressed teamwork and started to focus on processes rather than strictly on functions will have a leg up in a BPR effort.

Studying your company's readiness to change is one way of assessing what has worked in the past in terms of making change happen.

Throughout the duration of a BPR effort, it may be necessary to regroup because goals are not being met or a project is drifting off course. It is especially important at such a point to build from the successes rather than tearing down the entire BPR infrastructure and starting from scratch. Paying attention to what has worked is important in these "restarts."

Chevron: Linking Teams, Framework, and
Consultants Effectively

In 1989 Chevron Corporation, headquartered in San Francisco, set a five-year goal that board Chairman Kennth T. Derr articulated as a "return on stockholders' investment that exceeds the performance of our strongest competitors." To do this, Chevron embarked on an aggressive effort that included streamlining operations, improving work processes, and empowering employees by pushing decision making into the ranks closer to operations and closer to the customer.

By the end of fiscal 1993, Chevron had achieved a stockholder return of 18.9 percent, 5 percentage points greater than the average of its five chief competitors. The improvements were not always steady, and some of the efforts had to be reconstituted in midcourse.

Efforts were undertaken in all of Chevron Corporation's operating subsidiaries. Chevron Chemical Company (CCC), whose BPR project is discussed here, represents approximately 10 percent of Chevron Corporation's total revenues—$3.3 billion of $37 billion in 1993.

The efforts undertaken at Chevron Chemical Company provide a good example of how to constitute and reconstitute teams for a BPR effort that are different from the kinds of the teams or project groups a company is used to working with. It also shows the importance of using a framework to guide the teams' efforts and of how to use consultants effectively.

The chemical company has three divisions: Aromatics, Olefins, and Oronite. A fourth division, the Ortho division, producer of home and garden chemicals, was sold in 1993 as part of Chevron Corporation's efforts to divest noncore businesses. There are plants in 12 states, as well as in France, Brazil, and Japan. With five foreign affiliates and 18 subsidiaries, the company operates or markets in 80 countries.

Chevron Chemical, like the other Chevron subsidiaries, had its share of ups and downs during the five years of effort through 1993. But it found that stumbling is okay and part of the learning curve for many successful BPR efforts. However, executives did not allow obstacles to remain in the way of reaching their goals, and they were willing to take a few steps back to make giant leaps forward.

While many companies would have stopped an effort that touched critical points of customer contact when it threatened to go over budget and to fall behind schedule, Chevron executives regrouped and pushed forward with a new project manager and a new set of best practice approaches regarding teamwork, adherence to a framework, and aggressively managing and making the best use of outside consultants.

The BPR Approach

Chevron Chemical Company's first BPR project began in early 1992. A benchmarking exercise against the rest of the industry showed clearly that the company needed to improve how it delivered products to customers. In addition, the company's "command control" system for this and other key operations information was 25 years old and obviously out of date, complex, and slow.

For instance, during an as-is fact-finding mission to identify key processes for improvement, the reengineering team found that approximately 20 percent of the company's invoices were still being processed by hand. In addition, many of the information systems for each of the four divisions were separate.

Concurrent with a new financial and operations software package being chosen, the company embarked on a BPR effort to create a new customer order process that would result in one standard operating procedure for all four divisions. The focus was on significantly improved customer satisfaction and reducing cost of operations.

The project was organized in classic BPR team fashion, with a steering committee and project teams. The steering committee included a high-level manager from each division, the information systems department project manager, and outside consultants (one for strategy and project management expertise, and another for software systems expertise).

By June, the project was foundering, drifting behind schedule, and over budget. It was clear to the steering committee members that they needed to get the project back under control soon or risk losing support from the top and from the ranks. Jim Yochim, a seasoned project manager from the company's Houston office, was asked to manage the BPR effort.

Yochim's message to everyone involved in BPR—from steering committee members to reengineering work team members to consultants—was clear: He was not afraid to make changes. He conducted a complete project review, interviewed team members, and assessed objectives, work plans, and schedules. He evaluated the skill sets of existing participants against the project needs.

Initially, project teams had been formed around each major process to be reengineered. Chevron people were assigned full time. One or more consultants were assigned to each team. Each project team had coleaders, one a Chevron person and one from an outside consulting firm. In addition, a representative from the company providing the new software was on each team, as needed.

Each process team met once a month. A core team in each operating division was kept apprised of each process team's work. The core teams were chartered to provide specific information requested by the project team.

A senior manager was assigned to each project team as the "champion," responsible for trying to achieve buy-in of the project team's recommendations across the four divisions.

Making a Fresh Start

The changes that Yochim implemented included:

Replacing some team members and several outside consultants (including one of the senior consulting team members).

Adding team leaders from Chevron with business backgrounds (as opposed to IT specialists).

Redefining the project scope and insisting that the new scope be adhered to.

Utilizing project management software to manage the workload.

"I demanded the best and brightest," Yochim says. "And I wanted at least a 50-percent time commitment. You need the 'A team' to get to the right facts and develop the best solutions. And you need their time commitment to get their highest priority and keep the work moving."

The reconstituted reengineering project teams again formed around each major process selected for reengineering. But each team now had a coleader with a business background and a coleader/facilitator from the outside.

Teams worked to develop a comprehensive knowledge of the as-is for their process. This work became the basis for determining the performance baseline from which gaps and opportunities could be assessed, and gains from the new process measured.

"You have to control the project scope," Yochim says. "It was easy to drift slightly and add a little more each time, which was part of the problem earlier. Each time a problem was documented, management (or the team) wanted to solve it, which added to the scope. The real danger was not being able to get anything done."

Another important goal of the project restructuring was to create a sense of ownership of the processes and responsibilities for new solutions by reengineering work team members.

Each team was given objectives and target completion dates. The teams developed their own project objective statements, broke down areas or work into activities, and assigned responsibility to team members. A project management assistant kept track of teams' inputs, outputs, and schedules. This allowed the project manager to develop critical path, resource allocation, and deliverable information, and to generate management reports on the effort's status, which in turn enhanced executive buy-in.

Although all this effort took a lot of time and perhaps became too detailed (the process teams initially identified over 100 processes for potential reengineering), the data told a different story from the perceptions management had about the business. Based on the as-is data gathering done by teams, a filtering of potential processes for reengineering was done based on:

Impact on the organization.

Value as perceived by the customer.

Cost and benefits to the company.

The reengineering effort produced new processes for the company, complete with key metrics to maintain and monitor the processes, as well as skills/training needs assessments for people who would work in the new processes.

Using a Framework and Consultants

Chevron Chemical Company found tremendous value in using a structured framework; in its case, the outside consulting firm's framework was used.

"The framework helped us lay out what needed to be done, which gave us confidence that the teams would surround all of the issues, with nothing falling through the cracks," Yochim says.

A proven track record was important in selecting the right consultant. What Chevron found is that that track record is often embedded in an outside consultant's framework as it develops over time through working with a variety of consultants in many different industries.

Yochim believes the consultant and the framework both helped Chevron "get insights on the implications of what we were doing," and guided the company as it "focused on the critical 'must do' areas."

Yochim thinks this would have been very difficult to do without a lot of experience with previous reengineering efforts. The "right" consultant, Yochim believes, is:

[O]ne who can work together with your people as part of the team, and who has a proven approach. Company-only teams can't do all the BPR work without falling into old paradigms.

Consultants give another perspective and challenge the team to think out of the box for dramatic improvement. Without outside influence, it's hard to get the teams to stretch beyond the range of continuous improvement. If your consulting partners have been through this before, your chances of success are greater. These projects are risky, and it helps to be working with experienced partners.

FINDING THE *RIGHT* PROCESS

Best Practice 8: Link goals to corporate strategy.

Best Practice 9: Listen to the "voice of the customer."

Best Practice 10: Select the right processes for reengineering.

Best Practice 11: Maintain focus: Don't try to reengineer too many processes.

Prospective Best Practice: Create an explicit vision of each process to be reengineered.

We call the initial phase of a Business Process Reengineering effort the "Discover" phase. During this period—usually eight to 12 weeks—companies engage the BPR effort in a variety of ways. Most of the work in this phase of a BPR effort is undertaken by a high-level work team under the direction of the steering committee.

As we have said, work is often done through a series of all-day team meetings, sometimes called workshops. These all-day meetings are organized to create a high-level vision of what the organization can be like in the future, then to select a process or processes to reengineer. Meetings are facilitated to conclude with a planned outcome and to assign tasks to the team for completion by the next all-day meeting.

There are six major milestones:

Determine the company's strategic direction.

Assess the current processes (understanding the as-is).

Establish competitive baselining.

Find the "voice of the customer."

Select processes to reengineer.

Create process vision.

Figure 4-1 shows the sequence of these planning sessions.

At the same time the team is looking for reengineering opportunities, it is also uncovering short-term improvement opportunities, the "quick hits" that are so often crucial in gaining and maintaining support throughout the organization for the rigorous long-term effort. Gains from these short-term improvements are often used to "fund" the longer-range efforts.

In addition, the company's culture is analyzed during this early phase, and possible barriers to change are identified. The team must begin to develop effective strategies for overcoming those barriers. Development of a communication plan is also critical in this early phase; the right media need to be chosen and the basic messages are developed.

It is vitally important to understand that *doing a process right is not enough.*

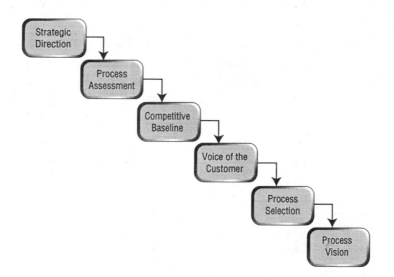

Figure 4-1. *The planning session sequence.*

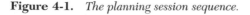

Picking the right *process is the key.* This is one of the major differences between a BPR effort and a TQM effort. In TQM the goal is to improve processes and functions throughout the organization in a step fashion. So which process is first out of the box is not crucial in a business competition sense, although it may be crucial in terms of gaining buy-in by choosing more "do-able" processes first. But, since the goal of BPR is to make a quantum leap in competitiveness—a high-risk, high-reward proposition—it is crucial to choose the right process.

While the goals of TQM are continous improvement throughout the company's operations, the goals of a BPR effort must be much more closely linked to the company's strategy for future competitiveness. Hence our first best practice in this regard.

BEST PRACTICE 8: LINK GOALS TO CORPORATE STRATEGY.

Corporate strategy must be the starting point for a reengineering effort because a strategic to-be vision gives the company a consistent course. Strategic planning activities often reveal the need for dramatic change and may even immediately pinpoint the processes that need transformation. Companies often put in a considerable effort early on to understand what drives competitive advantage in a particular industry.

The company's strategy is derived by asking a series of questions:

What is the product or service the company will offer?

Where will that product or service be offered, in terms of segmentation?

These are the standard strategist's questions. But then the focus turns to processes, as corporate executives ask:

What are the key processes (both core business processes and key supporting processes) that support the product or service in the related market segments?

What are the company's special competencies? How do those competencies enhance the company's key processes?

However the need for a BPR project is identified, its scope and direction need to support and link with the corporate vision of the future. Most organizations have a business strategy. Before BPR can be fully engaged, you must determine that your business strategy is:

Explicit.

Forward looking.

Well understood throughout the organization.

Viable given the current and future market conditions of the industry you are competing in.

To be a useful starting point for BPR, this business strategy needs to be based on:

A thorough environmental and competitor analysis.

A comprehensive understanding of the needs and capabilities of customer and suppliers.

An objective analysis of the organization's resources and capabilities.

Quantitative information that allows modeling of the revenue, cost, and resource implications of strategic alternatives.

Widespread senior management consensus on the strategic vision.

The company's strategy should also include objectives in specific competitive dimensions, such as:

Better meeting customer needs (attractive product features, service, price/positioning).

Superior economics (lower basic product/service costs, cost-to-serve advantage).

Time (more timely delivery, faster new product development).

Instead of relying solely on traditional structured methods of competitive positioning, such as market segmentation or product positioning, strategic assessment for BPR revisits customer and shareholder expectations, market dynamics, the role of information, and the core capabilities of the business.

CREATING A CONTEXT FOR STRATEGY

If major aspects of the strategic plan are missing, filling the gaps is an important first step. A clear understanding of what drives competitive advantage and where the company wants and needs to go are critical elements of BPR discussions. This understanding is developed through:

Customer research.

Competitive analysis.

Benchmarking.

Financial review.

Operational review.

Information management review.

Assessment of key performance indicators.

Organizational culture assessment.

BPR involves an intense focus on *customers,* discussed in greater detail in the next best practice.

Market research, through surveys or focus groups, provides an external vision of quality and value. This information is critical to giving BPR projects the capability to produce the right kind of dramatic change.

Not just any customer will do for this type of assessment. Most organizations base their market research on the broad, undifferentiated mass of customers. But we believe you should focus customer research in this first phase of BPR on what we call your "barometric" customers—those who are on the leading edge, with the vision to respond to innovation and anticipated change. When you look for breakthroughs and new ways of doing business, their opinions are most important.

While significant process improvement can result from focusing on internal conditions, you can't catch or overtake the competition unless you know where you are relative to them—unless you engage in *competitive analysis*. Detailed knowledge of competitors' strengths and weaknesses can help identify where your company's competitive edge may lie.

To complete the picture, you also need to understand the dynamics of your industry and market. Companies get information on the competition and market in several ways. They survey their competitors' customers to get their perspectives on strengths and weaknesses. Trade associations, survey research firms, chambers of commerce, and advisory panels can also provide information.

Astute organizations also use their sales force to gather intelligence on what competitors are marketing. Often, this is a company's "first alert" to rival products and services.

Benchmarking is another essential tool for strategically driven BPR. Benchmarking may involve comparisons at three levels:

1. Internally, comparing performance of similar units or divisions.

2. Within an industry.

3. With the "best-in-class" performance of a company in any industry.

Figure 4-2 shows a hypothetical benchmarking exercise.

David Kearns, former CEO of Xerox, calls benchmarking "the continuous process of measuring product, service and practices

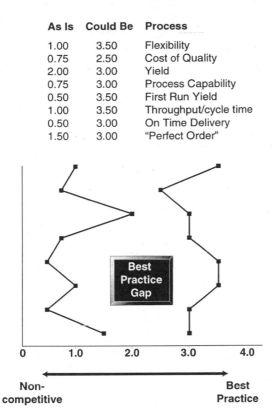

As Is	Could Be	Process
1.00	3.50	Flexibility
0.75	2.50	Cost of Quality
2.00	3.00	Yield
0.75	3.00	Process Capability
0.50	3.50	First Run Yield
1.00	3.50	Throughput/cycle time
0.50	3.00	On Time Delivery
1.50	3.00	"Perfect Order"

Figure 4-2. *A hypothetical benchmarking exercise.*

against the toughest competitors, or those companies recognized as industry leaders."

Benchmarking is essential to a company's strategic vision in that it helps you understand the "what could be" of your future state. The flip side is doing an objective appraisal of your company's capabilities. Then, by measuring the gap between your present capabilities and the best in class across many functions and processes, you can set priorities for the processes to reengineer.

While it's important not to limit creative vision in BPR, reengineering must be reality-based. You need to understand the critical barriers to success within your operations, your supplier network, and your corporate culture that will need to be breached in order to successfully implement your reengineered process(es).

You need to assess your current capabilities across five dimensions, in order to define the implications that process change will have on the current business:

A *financial review* analyzes financial and management accounts to identify and explain significant trends and implications, as well as to define an operational base case. A financial review asks the questions:

Is the business reaching its financial goals?

Which processes are the critical levers?

How can they be used more effectively to create profitable growth for the company? (Or, in a declining business, how can they be used more effectively to create profitability for the company?)

An *operational review* assesses and explains the company's operational effectiveness and efficiency. This gives you a first look at your current process capabilities. This review answers the questions:

What can the organization do, and what does it do especially well?

Which processes need enhancement, and which can be leveraged?

An *information management review* assesses the current and future roles of information technology services (ITS) and the extent to which they can provide support to core business processes. This answers the questions:

Can executives see, in real time, what is happening across the business?

Does the IT system allow processes to run effectively and efficiently?

Are there any ways in which processes are constrained by the current IT system?

Key performance indicators are used in monitoring core processes and organizational behavior. The questions these indicators seek to answer are:

How is the business doing?

How do executives monitor the business?

How should executives monitor the business? Remember the old adage, what gets measured gets attention.

A *cultural assessment* reviews the aspects of an organization's culture that will promote or resist the major changes brought about by BPR. This assessment gets at the questions:

Can the organization change, and how easily can it change?

What are the barriers to change?

What is the level of resistence to change?

All our survey respondents said their BPR goals were closely linked to their corporate strategy. The goals and strategy were almost uniformly customer-focused (68 percent of respondents) or quality-focused (45 percent of respondents), as shown in Fig. 4-3. Some examples:

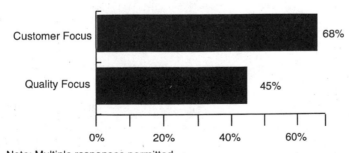

How were your BPR goals linked to corporate strategy?

Customer Focus — 68%

Quality Focus — 45%

0%   20%   40%   60%

Note: Multiple responses permitted.

Figure 4-3. *Results of survey of BPR goals.*

Our strategy is to improve service while decreasing cost.

Our goal was to focus on quality and service, while reducing all nonvalue-added activities.

MAPPING THE CORE BUSINESS PROCESSES

As part of the operational review for this strategy alignment, you need to make a first assessment of the core business processes and key supporting processes. It is most likely that your choices for processes to reengineer will come from these two groups, and, if possible, from the ranks of core business processes.

When you determine these processes, you need to create what we call a quick map. A *quick map* is a first-cut representation of a process—essentially a rough diagram of the boundaries, external connections, and process flows for each process. For simple processes, the quick map may begin to point you in the direction of how to reengineer the process. For more complex processes, you will need to dive down deeper into the process to find sub-processes that become targets for reengineering.

For instance, the core business process of new product design in the automotive industry is far too large to reengineer as a whole; a more manageable effort might be the design of a new antilock braking system or a new drive train. When the U.S. Postal Service sought to reengineer its process for delivering Express Mail, it quickly determined that the process was too big to be tackled at once. It refined the effort and focused on reengineering the subprocesses of account management, billing, tracking, and other forms of support, leaving operational subprocesses such as receiving, sorting, and delivering for a future effort.

On the face of it, the processes chosen for reengineering were back-office, noncore support processes. When most of us think about mail processes that "touch" customers, we think of the workers who handle mail drop-off or mail delivery. But for

Express Mail, the Postal Service is trying to compete against overnight package companies for large, corporate customers. For these customers, billing, tracking and tracing, setting up new accounts, and a host of other subprocesses are key.

C A S E   S T U D Y

Marion Merrell Dow: Rethinking Prescription Drug Delivery to the Marketplace

A revolution in the U.S. health care delivery system has been underway since the mid-1980s. Primarily propelled by economic consideration, the pace of change accelerated in the 1990s as health care reform became a subject of debate and as consolidations within the industry began to take place.

From the early 1950s to the early 1990s, health care in general and the pharmaceutical industry in particular were an ever growing sector of the U.S. economy. Research efforts brought out a succession of "blockbuster" drugs for the treatment and management of a number of diseases that had previously required surgery and long hospital stays.

Health care costs also rose astronomically through that period. By the late 1980s and into the 1990s, government and especially corporate buyers of health insurance were scrutinizing health care cost to a far greater degree than before. Payers began to press for less expensive means to deliver care, as well as controls to assure that economic considerations were taken into account when surgical, radiation, and pharmaceutical treatments were prescribed.

At Marion Merrell Dow, and throughout the pharmaceutical industry, the 1990s brought a new era. No longer were physicians unfettered in their therapeutic practice. No longer would the traditional techniques used to promote pharmaceutical products be effective in garnering revenues.

Marion Merrell Dow, Inc., headquartered in Kansas City, is a company that came about in response to changing market conditions. The 1989 combination of Marion Laboratories with Merrell Dow Pharmaceuticals was one of the first mergers of large pharmaceutical companies in what by 1994 would become a wave of such mergers.

Marion Merrell Dow's products are marketed through programs targeted to reach physicians and other health care professionals, managed care providers, and hospitals and other institutions. Branded prescription products are marketed directly by the company while over-the-counter products are marketed through SmithKline Beecham Consumer Healthcare.

Though the bulk of the company's revenues are generated in the United States (64 percent), almost 20 percent comes from Europe and the remainder from a host of other countries, including Canada, Japan, Australia, and New Zealand.

1992: A Business Case Is Made

The United States Business Reengineering Project (USBR) was broached in a conversation in December 1991 between the company's chairman and CEO, Fred W. Lyons, Jr., and John L. Aitken, Ph.D., the company's vice president for quality performance improvement.

Since the company's inception in 1989, an effort had been underway to inculcate TQM principles throughout the organization. Training and other techniques had been used to develop awareness of TQM. A major process improvement effort involving the company's R&D organization had been conducted, as had a number of smaller initiatives focused on specific business processes, such as procurement and clinical data management.

In Lyons' view, the time had come to apply these techniques to a major strategic challenge, namely reengineering the company's entire commercial approach to fit the "revolutionary" changes taking place in its largest market, the United States.

During the early months of 1992 Aitken, the project's chief architect, put in place the four key components of the project, which provided a framework for the project's conduct and remained more or less intact through the end of 1994:

1. A steering committee of senior functional executives, headed by Peter Ladell, the company's North American president.
2. A cross-functional project team comprised of some of the company's best and brightest associates, to guide the project through the first three phases.
3. A consultant/facilitator for the reenginering effort.
4. An ongoing "customer-based, customer research" effort, run by an outside market research firm.

In addition to serving as the project's sponsor, the steering committee served to provide "political cover" for a project that, given the degree of change expected, was bound to encounter resistance and barriers.

The project charter, defined by the steering committee, was:

To develop and implement business processes that successfully meet the changing needs of U.S. customers while effectively meeting or exceeding ongoing business goals.

The project timetable was envisioned as:

Stage 1: Current State Documentation (6/92–9/92)

Stage 2: Benchmarking (9/92-11/92)

Stage 3: Visioning (12/92)

Stage 4: Future State Design (1/93–6/93)

Stage 5: Implementation (7/93–9/94)

Stage 6: Measurement and Continuous Improvement (ongoing)

Current state documentation was based on a process model of Marion Merrell Dow's commercial activities, created by the project team. In retrospect, this was little more than a classic functional model "turned on its side." The notion of thinking in process terms was new and difficult.

As simple as the model was, however, it served to provide a foundation on which cross-functional teams were built, consisting of associates who were familiar with one or another of the activities involved in each process.

More than 100 associates prepared detailed process descriptions, setting forth the ways in which Marion Merrell Dow then handled these major processes. Process flow diagrams were constructed, as well as forms setting down the "inputs, transforming activities, and ouputs that occurred at each step of a process."

While the current state documentation was an often tedious exercise in paperwork, it provided the project team a clear sense of what was working and what wasn't. It also highlighted the ways resources had been allocated and the ways they might have been allocated for satisfying the needs that customers said were most important to them.

Benchmarking was initially undertaken as "secondary research," using electronic information networks to search for published information describing what—on the surface—appeared to be better performing processes similar to those that needed reengineering at Marion Merrell Dow. This research turned up a number of companies whose processes appeared to be more effective than Marion Merrell Dow's. Fortunately for the company, none was in the pharmaceutical industry.

Through networking, the project team was able to visit a number of these companies, including Hewlett-Packard, AT&T, General Electric, USAA, and Federal Express. The insight gained from the intensive and thoroughly planned site visits proved to be of great value as the team engaged in the visioning process.

Another important, if untraditional, component of benchmarking during the USBR project involved the "total immersion" of the project team in days and days of exposure to industry experts from both inside and outside the company, including those who had made studies of expected health care legislation, developments within the pharmaceutical industry, and the broader issues in health care, including organizational design and information technology.

After completing the benchmarking, the team went off site for a week to engage in an exercise to create a new *vision* for Marion Merrell Dow.

The vision and a new process model that complemented the vision were presented to, and approved by, the steering committee at an all-day meeting in January 1993. The overall theme of both the vision and the new process model is that Marion Merrell Dow must become a customer-centered company, and that by doing so the company can make sure that all patients who might benefit from its products have access to them.

BEST PRACTICE 9: LISTEN TO THE "VOICE OF THE CUSTOMER."

Because core business processes by definition connect with customers—as opposed to supporting processes, which can be completely internal—it is important to obtain meaningful customer input before deciding which processes to reengineer. We call this listening to the "voice of the customer."

You can obtain customer input in a number of ways. It's important to use techniques that allow you to actually hear customers articulate reasoning for the decisions they make and for the needs and desires they profess.

Eighty-two percent of our survey respondents conducted customer surveys. Additionally, 31 percent conducted focus groups and 14 percent went on site visits, as shown in Fig. 4-4. Fourteen percent even had customers on their BPR teams.

We believe a good customer survey—whether one-on-one or by means of a focus group—should include questions about the customer's requirements for the supplied good or service with regard to a host of attributes such as quality, cost, delivery, reliability, and after-sales service and support. Customers should be asked to rank order competitors against the attributes. In this way, you can see what the customer base perceives to be your strengths and weaknesses.

Equally important, questions should be asked as to what we call the market parity requirements both today and in the future:

What does a competitor need to be able to do for the customer today to be competitive?

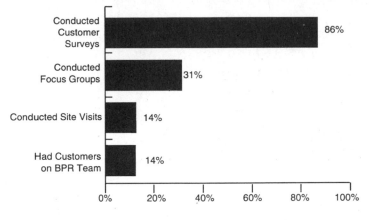

How was the voice of the customer taken into account in the redesign?

Conducted Customer Surveys — 86%

Conducted Focus Groups — 31%

Conducted Site Visits — 14%

Had Customers on BPR Team — 14%

Note: Multiple responses permitted.

Figure 4-4. *Use of customer information.*

What will a competitor need to be able to do for a customer two, three, or five years from now to remain competitive?

The answers to those questions tell you the "get right" requirements you have to think about as you seek to reengineer your core processes.

This exercise may also cause you to redefine your core business processes according to where the customer base puts its emphasis. For instance, if a significant number of customers say they would like to see shorter lead times, in the future order entry may be defined as a core business process, even though in the past you have defined it as a supporting process.

At this point you can compare your quick map and first-pass understanding of your process capabilities against what customers say is important today and will be important tomorrow. From this knowledge you can create what we call a *best practice vs. can-do* map that shows your current capability against the industry best practice mapped for all of the attributes that customers describe as important. Such a map is shown in Fig. 4-5.

Breakpoint Priorities

	LOW	HIGH	
Time		✕	Potential Breakpoint
Quality	✕		Unlikely Breakpoint but Keep Right
Cost		✕	Get Right
Reliability		✕	Evolving Breakpoint
Process Design		✕	Potentially Emerging Breakpoint
Flexibility		✕	Get Right
Differentiation Optionals	✕		Keep Right

 Industry Best Practice

✕   Your Current or Potential Capability

Figure 4-5. *"Best practice vs. can-do" map.*

Marion Merrell Dow Listens to the Voice of the Customer

The "customer-based, customer research" Marion Merrell Dow commissioned from the Anjoy group was accomplished in two stages.

In the first stage, focus groups of physicians, pharmacists, health care administrators, regulators, and members of other key customer elements were conducted to identify the issues and concerns important to customers as they thought about dealing with a pharmaceutical company. In this way, *customers defined the key topics for the subsequent research.*

Marion Merrell Dow developed a customer "roles" model, which is made up of the six key "players." In the world of pharmaceuticals, the array of customers is large, and only one is the individual who actually uses the product. The six customers are:

1. *Therapeutic beneficiaries:* For example, patients who consume pharmaceuticals products and their caretakers, or parents of children who are patients and caretakers of elderly patients.

2. *Clinical decision makers:* Those who prescribe and counsel the use of pharmaceutical products: physicians, pharmacists, and other health professionals.

3. *Economic decision makers:* Those who influence or control access to pharmaceutical products on economic grounds: health care administrators, insurers, and others.

4. *Payers:* Those who pay for pharmaceutical products consumed: corporations, insurers, governments, and patients.

5. *Public-policy makers:* Those who regulate or influence the availability of pharmaceutical products: the FDA, other government agencies, legislatures.

6. *The dispensing pipeline:* Those who deliver or influence the distribution of pharmaceutical products from the manufacturer to the dispensing point: wholesalers, chain pharmacies, mail order pharmacies, hospital pharmacy management companies, and so on.

In the second stage, in-depth telephone interviews were conducted with broad-based and statistically determined subcategories of the relevant customer categories (i.e., 250 interviews with physicians, 300 with pharmacists, 250 with managed care administrators).

In these interviews, the relative importance to the customers of the issues determined through the focus groups was discovered. In addition, participants were questioned about their satisfaction with the performance of the industry in general, Marion Merrell Dow in particular, and the "best" pharmaceutical company, across each issue.

From this work, Marion Merrell Dow received detailed measurements of its performance relative to the industry in general and to the top performing companies in each area. The company also received an appraisal of the value customers placed on the features and benefits of its products and services.

There were a couple of surprises for the company. One was that, in general, there is little differentiation by customers among major pharmaceutical companies. The only company that was rated highest on a statistically significant number of items was

Merck. Otherwise, all major drug companies received a highest rating on about the same number of items.

The second surprise, although not a big one, was that all the customer categories said that economic considerations are increasingly important. This seemed to contradict anecdotal information about continued conflict between payers and prescribers. As Aitken says:

It shouldn't be that surprising that one of the big unsatisfied needs customers have is principally economic. All American drug companies get high marks for safety and effectiveness, but increasingly the economic needs are as critical as the safety and effectiveness. Customers say drugs simply have to be less expensive.

Marion Merrell Dow also used this work to create a two-by-two matrix such as that shown in Fig. 4-6. Items that appear in the lower right-hand box pointed the way to issues that clearly needed improvement, while items in the upper right-hand box gave the company an idea about where its potential competitive breakthrough might be.

We'll discuss the company's process changes in the next section.

Marion Merrell Dow Does Very Well

Relatively Unimportant Needs of Customers	**Area for Strategic Reassessment**	**Potential Area of Competitive Advantage**	Relatively Important Needs of Customers
	Possible Area for Savings Opportunities	**Priority Areas to Reengineer**	

Marion Merrell Dow Does Not Do Well

Figure 4-6. *Two-by-two matrix of customer survey results.*

BEST PRACTICE 10: SELECT THE RIGHT PROCESSES FOR REENGINEERING.

Selecting the right process sounds difficult, almost ominous. However, in many instances the right process "selects itself" as you do your business strategy analysis and listen to your customers and potential customers. Sometimes the appropriate process all but jumps out of the best practice vs. can-do map. Other times you must search harder.

To fundamentally change work processes, we must first define them. Then we must understand where they fall along the continuum of strategic importance. Our strategic continuum has four categories.

1. The most important strategic processes are *identity processes*. These processes define the organization to itself, to customers, and to investors.

2. Next come *priority processes*. These directly and significantly affect everyday performance.

All a company's core business processes are either identity processes or priority processes. While it is desirable for a company to have an identity process, few do.

3. Lower still on the strategic hierarchy are *background processes*. These are necessary for the business to survive in the long term.

4. Finally come *mandated* processes, carried out due to government or other regulations. Most accounting processes are mandated.

To understand how to define processes and rank them along the strategic continuum, let's look at a fictional company called Teleservices. Teleservices' major processes are named and defined in Fig. 4-7. Next, Teleservices' major processes are placed along the strategic continuum in Fig. 4-8.

Figure 4-8 also cross references where the processes lie on the strategic continuum by how efficient the process is in its current state.

Process Name	Definition: Work involving...
Customer acquisition	...finding, qualifying, pitching, closing new customers
Customer migration	...developing new business with current customers
Order entry and fulfillment	...delivering on our bargain with a customer
Inquiry and complaint handling	...customer-initiated contact after sale
Local market opening	...a concerted response by geography

Figure 4-7. *Teleservices' major processes.*

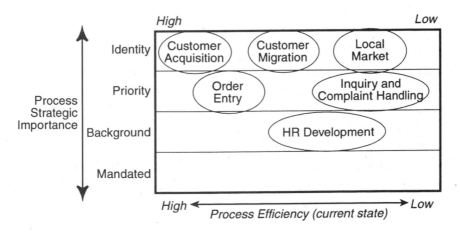

Figure 4-8. *Teleservices' major processes in terms of the strategic continuum.*

Marion Merrell Dow Chooses Processes

Marion Merrell Dow's two-by-two matrix served the purpose of a process efficiency rating, and its voice of the customer research helped it determine its identity and priority processes with regard to customer contact. The goal clearly was to find processes to reengineer in order to make delivery of branded pharmaceuticals (prescription drugs, whether on or off patent, identifiable by a corporate brand name) less expensive. In that way, the company could meet some of customers' economic needs without sacrificing funding for research and development, the life blood of all pharmaceutical companies. As Aitken puts it, "We didn't justify this project by saying it would generate one penny's worth of new sales, but that it would reduce expenses to reach the marketplace by $65 million over four years."

From there, the United States Business Reengineering (USBR) project was able to focus on the right processes to reengineer. The company identified 13 processes and has so far embarked on reengineering 11.

One very simple process to reengineer is in marketing collateral material. Between $20 million and $30 million in savings will come simply from a reduction in the amount of marketing material published and distributed. It simply doesn' t meet customer needs.

"While this material has been thoughtfully put together, artistically rendered, and is of a high professional caliber, it has been made to appeal to such a broad audience that it doesn't answer the specific questions customers have," Aitken says. "No doctor is going to wade through a six-page glossy brochure to find the one line about what side-effect the particular drug might have on a pregnant woman."

The process of sales representative interaction with individual physicians was changed. Traditionally, reps carried a car full of literature for doctors and giveaways (pens, pads, refrigerator magnets, and the like) for the doctors and their staff, and they had a garage full of such items at home. Sales reps now order daily for the customers they have seen—literature that answers specific questions—and for customers they will see the next day—give-aways the rep knows the customer and staff like. The rep sends this information to a marketing material distribution center using new technology acquired as part of the process change (a modem and a lap-top computer), choosing from the hard-disk-based catalog of over 400 items, ranging from pens to scientific journal articles. The material is delivered the next day to the customer's site.

Another process change involved pharmacists. Forty years ago many pharmacists, especially rural pharmacists, had trouble gaining enough continuing education credits for recertification. So pharmaceutical companies developed and ran continuing education programs. Today, practically every pharmacist in the United States has access to such continuing education through community colleges, and utilizes pharmaceutical company courses much less than in the past. "The whole process meets less than a compelling need," Aitken says.

Another change was more drastic, changing the way the company interacts with the increasingly powerful large customers. Rather than individual representatives selling to any organization within a given zip code, the customer base was broken up into account types, and account teams were developed to market to account types over a larger geographic area.

This meant completely altering the traditional field service organization structure. District managers no longer drive the field staff game plan; instead, account mangers drive the way account teams operate. The cultural changes are enormous.

Individual reps now "essentially have two bosses, the district manager and the account manager." District managers are being asked to assume a more "coaching, mentoring" and employee development role, while much more of the strategic decision making is falling to account managers, who are being asked to carve out a niche within a strong corporate culture.

Aitken conceded that such changes are difficult and that not everyone will be able to function in the new environment. Corporate leaders have recognized the need to install new metrics and new incentives for pay into the system, which Aitken hopes will ease the transition. But, as he says, "If we show success in making some changes within two years I will consider this a success."

But those kinds of changes will need to be made if Marion Merrell Dow is to continue being competitive in the marketplace, where economic decision makers are increasingly wresting power from therapeutic decision makers, and where drug companies are looking to develop long-term across-the-board relationships with large-volume buyers.

Across all the process changes, Aitken sees a fundamental shift in the way the Marion Merrell Dow sales and marketing staff needs to think about the work it does. "We need to have a shift in the understanding of our job from selling to educating our customer communities. Virtually every customer subcategory identified as a major need 'unbiased' information. That was the word they used."

From patients who take the drug, to doctors who prescribe it, to Health Maintenance Organization directors looking to create plans with a drug company for comprehensive delivery of pharmaceuticals to all plan beneficiaries, "customers are more savvy than ever at distinguishing between information that is unbiased, scientific and technically sound and information that is meant to influence their purchasing decision. We need to talk less about sales and marketing and more about providing benefits and value to our customers. We need to talk about being a partner in the health-care process."

BEST PRACTICE 11: MAINTAIN FOCUS: DON'T TRY TO REENGINEER TOO MANY PROCESSES.

The entire thrust of BPR is to conduct a rigorous analysis of which processes—usually core business processes but occasionally a key supporting process—will most cost-effectively increase your competitiveness.

Because it is highly risky to make major changes in these processes, it is imperative that an organization's energy be tightly focused on reengineering only a few processes.

We have always said that organizations should focus on three or fewer processes during each round of designing and implementing reengineered processes. Our survey results bore out that that advice is on target; companies focusing on one to three processes in their reengineering effort reported more satisfaction with their BPR than those reengineering more processes.

Fully 73 percent were reengineering one to three processes, as shown in Fig. 4-9. The largest percentage of companies—31 percent—were reengineering two processes.

Some survey respondents reported that, as the number of processes being reengineered increased, so did confusion over the goals and objectives of the entire effort, because "so much was going on." If the participants in project teams were confused, imagine what the average employee, who was not involved in a BPR team, was thinking. And how can a steering committee of executives focus on five, six, or even seven BPR projects when they also must spend time running the day-to-day operations?

As Lawrence Bossidy, CEO of AlliedSignal, says, BPR is "gut-wrenching stuff." And it must be executed in manageable doses.

How many processes did you reengineer?

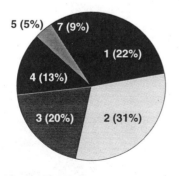

Figure 4-9. *Number of processes undergoing reengineering.*

PROSPECTIVE BEST PRACTICE: CREATE AN EXPLICIT VISION OF EACH PROCESS TO BE REENGINEERED.

A process vision must be developed for each process being considered for reengineering. This process vision characterizes the new process by its attributes, which in turn suggests how the process improvements will be carried out. Different process characteristics lend themselves to different solutions for process reengineering; for example, technological characteristics, such as expert systems or electronic data interchange (EDI), through functional consolidation, through concurrent engineering, using customer partnerships, and so on.

The idea of a vision starts when senior executives create a vision for the business, as it should be in the future. This is essentially defining the strategy: What will the company sell? To whom? What processes can the company leverage to do this?

Then process teams, with the help of the core BPR team that does the initial BPR analysis, create an explicit process vision for each process that will be reengineered.

Finally, the steering committee and the top corporate executives check the process visions for consistency with the overall corporate vision. This serves to synchronize executive thinking with team thinking. When senior executives examine a process vision, they must ask: Does this make dramatic changes or merely incremental changes? Is the pain that the company will encounter in this undertaking worth the gain that is envisioned?

The process vision describes:

The new process' capabilities, as well as the expected performance improvements in time, quality, cost, and service.

How the new process will support the strategy, respond to the customer, and respond to the competitive challenge.

A process vision must include objectives for the new process, including targets that illustrate dramatic improvement, such as a cycle time reduction of 50 percent, a reduction in processing time of 60 percent, or the like.

A matrix of objectives and attributes, such as the one in Fig. 4-10, is helpful for presentations to the executive steering committee.

Only when there is an explicit process vision based on fact-driven analysis can the executive steering committee give the explicit approval necessary for the reengineering effort to move from the planning phase to the phase of actually reengineering the chosen processes.

Such fact-driven analysis comes from rigorous research and analysis of the as-is state of the organization and the major business processes, and only after listening to the voice of the customer. As John Aitken, who headed up the U.S. Business Reengineering effort for Marion Merrell Dow, says:

> Having the data is tremendously valuable when you go to advocate the need to make substantial and dramatic changes. I had customer data no one could argue with. Managers' opinions were no longer relevant. If someone said, "but when I was a rep, or when I was a district manager, …" I could just point to the data about how different the world is today.

Process Objectives	Process Attributes
Quantified targets for change: Cycle time reduction of 50% Reduce customer processing time and cost by 60% Double customer service satisfaction measures	Characteristics and enablers: Use of EDI to link with customer Expert systems for credit checking Automated proposal generation Key activities performed by empowered teams Use case management approach for customer interface

Figure 4-10. *Matrix of objectives and attributes.*

DOING THE RIGHT PROCESS RIGHT

Best Practice 12: Maintain teams as the key vehicle for change.

Best Practice 13: Quickly come to an as-is understanding of the processes to be reengineered.

Best Practice 14: Choose and use the right metrics.

Prospective Best Practice: Create an environment conducive to creativity and innovation.

Prospective Best Practice: Take advantage of modeling and simulation tools.

N ow that you have found the right process—or more likely the right two or three processes—it's time to reengineer them so that you do the process right.

A separate team for each process needs to be established and chartered with reengineering the selected process. The key question each team must answer is, What should the process look like in the future?

That question has within it four component questions:

1. What will the process do?
2. How will the process be done?
3. How will we know if we are doing it right?
4. How do we make the transformation in the organization necessary to make the process change stick?

This phase of BPR involves all the skills and management abilities inherent in the design of any physical product or any service. The selected process is taken apart and conceptually put back together in a better way. A specification of the reengineered process is prepared that includes new ways of working, a supporting organization structure, requirements for information systems, and the appropriate performance measures.

REFOCUS COMMUNICATION AND VISION

Communication needs to be reexamined and refocused. Changes that have already occurred in the organization and in the processes need to be identified and communicated throughout the company. For the next round of communication, the "best fit" media need to be identified, as do the critical constituencies.

The focus should be on communicating the "right" message at the "right" time to the "right" audience segment in the organization or in the process teams. Much of this communicating should be done by a powerful spokesperson, the reengineering champion whenever possible.

In the same way that a two-dimensional representation of the three phases of BPR—discover, reengineer, implement—does not truly depict how the effort is physically carried out, neither does any two-dimensional representation of the steps within any one phase. Many of these activities, in fact, spill over from one phase to another.

Such is the case with creating a vision. By combining an understanding of basic process capabilities with the voice of the customer information, the executive steering committee extends its vision of the business in its to-be state. This vision is tied to customer expectations and drives the selection of processes to be reengineered.

This macrovision of the business is passed to all the individual process teams, each of whom then creates a vision of the individual process they are working to reengineer. Each team defines the to-be of the individual process in terms of:

The skills necessary.

The activities that will be undertaken.

The information and information technology necessary.

The behaviors necessary to work in the new way.

The way the process, and the behavior of individuals, will be measured.

While you are working to reengineer the various processes for breakthrough results, you will find subprocesses or components of processes in which significant improvements can be made without a full-scale reengineering. Some efforts to improve these areas should be taken; these improvements can "fund" more dramatic efforts in other areas of the process.

But don't fall into the trap of only going after the low-hanging fruit and calling that reengineering; it is not. If you have organized for and motivated the organization for reengineering, and then settle for process improvement, there can be severe disappointment, among both employees and shareholders.

Even if senior corporate executives are championing the effort, before moving from the reengineering phase to full implementation, it will be necessary to meet with the executive steering committee to obtain a formal go-ahead for implementation.

This phase of a BPR effort is a period of transition and fraught with danger. The best practices we have identified reflect the need to maintain focus and continuity, at the same time setting up the foundation for the profound changes that will occur within the organization during implementation of the reengineered processes.

BEST PRACTICE 12: MAINTAIN TEAMS AS THE KEY VEHICLE FOR CHANGE.

It is usual at this point to reconstitute BPR teams to some degree or to add teams. Much of the work of defining current processes and making a preliminary decision about what processes to reengineer was accomplished at the level of the executive steering committee, with small teams created to define processes or educate the steering committee.

Once the processes to be reengineered have been chosen, it is time for BPR teams well versed in the particular processes to roll up their sleeves and go to work. The membership of these reengineering teams needs to be cross-functional; the chosen members should work in the most important parts of the process, and among them should be all the necessary skills and backgrounds needed for the project.

Wise organizations choose the process owner, the executive or senior manager who will be in charge of the process after reengineering, as the team sponsor. This assures the owner's total commitment to superior results.

In our survey, almost 60 percent of respondents said their teams changed membership throughout the BPR effort. In one-third of the cases, new skill sets were needed, which necessitated bringing new members on board. In about half the cases, tasks were completed, which meant team members with particular skill sets could leave to pursue other activities.

This ability to have fluid, dynamic teams is made necessary by the fact that, while individuals are serving on teams, they are often devoting 75 percent to 100 percent of their time and effort to the BPR activities.

Of our survey participants, 50 percent of new team members were appointed, while about 33 percent volunteered for the assignment. See Fig. 5-1.

Several respondents mentioned that problems occurred when team members changed, and stressed the importance of building overlap and team continuity into the team selection process throughout the BPR effort. One team cannot be disbanded and another started at this phase; the new team must contain members from the initial team that did the preliminary work.

Changing Team Members: Maintain Continuity
Almost 60% of the respondents said that their teams changed throughout the project. Reasons for the change focused on: – New skill sets needed in one third of the cases. – Completion of the tasks in half the cases. New members were appointed 50% of the time and were voluntary 33% of the time. Several respondents mentioned that problems occurred with team member changes and stressed the importance that overlap and continuity be built into the team selection process throughout the project.

Figure 5-1. *Results of survey on fluid, dynamic teams.*

In our survey, we found that senior line management such as division heads were always on the teams, often the main points of continuity. These division heads were the sponsors of the BPR effort in about 42 percent of cases; CEOs were the champions in 44 percent of cases. Figure 5-2 shows the multitude of primary sponsors.

In reengineering, just as in product design, many development tasks are managed in parallel. This means that team members can address concurrently many different parts of a reengineering effort, such as information technology, work flows, and human resources. This approach reduces reengineering lead time and costs. If done well, it also mitigates against the kind of hand-off errors and miscommunication inherent in sequential development.

Team Building and Team Training

The diversity of team members and the need for parallel development require superior teamwork; achieving this is the first task of a team that engages in BPR. Usually led by an experienced facilitator/instructor, members "build" their team by clarifying goals

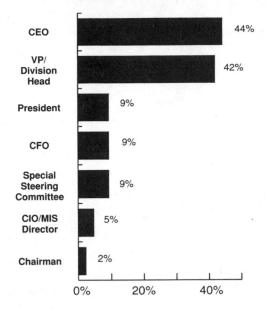

Who was the primary sponsor of the BPR project?

CEO	44%
VP/ Division Head	42%
President	9%
CFO	9%
Special Steering Committee	9%
CIO/MIS Director	5%
Chairman	2%

0%   20%   40%

Note: Multiple responses permitted.

Figure 5-2. *Range of primary sponsors.*

and expectations and by agreeing on a set of rules for team conduct. Through team-building exercises, they develop a team dynamic conducive to cooperation and synergy. The ultimate outcome is agreement on the procedures the team will follow in its work, which should include reporting and control, tools and techniques, and roles and responsibilities.

Since the team often uses tools and techniques that are new to its members, an instructor/facilitator leads members through the initial application of these methods to real life issues that are part of the reengineering effort. This as-needed mode of training is more conducive to adult-style learning than is rote instruction. Often, the facilitator will remain with the team over several weeks or months, coaching the members in tools and methods and facilitating meetings.

Why do all this? Because the reengineering team is your single most valuable resource in a BPR effort. A poorly coordinated team will result in major delays and problems. Investing in good teamwork is money well spent.

BEST PRACTICE 13: QUICKLY COME TO AN AS-IS UNDERSTANDING OF THE PROCESSES TO BE REENGINEERED.

This is the central information-gathering step in the reengineering phase of the BPR effort. In the early phase, a high-level process map—a quick map—was done of all major business processes to help select processes for reengineering. In this phase, it is important to get a more detailed understanding of the few processes that have been chosen for reengineering.

Getting a solid view of the as-is state is important for a number of reasons. It allows you to:

Keep the parts of the process that are important.

Create a migration plan.

Know who is involved in the process.

Create a common understanding for people.

Create a fact-based performance baseline.

Find the internal controls in the current process, so that, when the process is reengineered, appropriate controls are provided. Figure 5-3 shows the importance of process mapping.

It is important to understand the process boundaries, what is in the process, and what is outside. These boundaries are determined by the top executives, who then give directions to the teams.

If design parameters were not set by the steering committee in the early phase of reengineering, they are set now. These basic

Why "As-Is" Process Mapping?
Develop a common, fact based understanding of the current process among all team members.
Allows team to "keep" any of the best parts of the process.
Ensures important internal control functions are not completely overlooked in the design of the new process.
Identifies who is affected by or involved in the process and thus identifies where change management efforts should focus.
Facilitates development of a migration plan.

Figure 5-3. *The importance of process mapping.*

design parameters, which will govern your decisions about reengineering alternatives, may include:

Centralizing (or decentralizing) operations whenever possible.

Following a certain process model (i.e., the flexible factory or Just-in-Time cycle time reduction).

Shifting responsibility for some types of tasks to customers or suppliers.

Providing easy information access to employees, managers, customers, suppliers, and other members of the enterprise.

Minimizing supervision.

Outsourcing noncritical activities when it is cost-effective to do so.

Tying performance measures to customer or shareholder satisfaction.

Using information technology to guide employee decision making.

Providing a single point of contact for customers or suppliers.

Building continuous improvement into all processes.

The challenge is to find a set of design parameters that meet your company's individual needs and that are inherently consistent with one another.

Knowing your design criteria will help you determine how detailed you want your process mapping to be. Process maps should be detailed enough to break a process down into more manageable units for teams to work on redesigning. Even if many activities within the as-is process are to be eliminated in the reengineering, it is important that the map identifies those activities.

Without this level of detail, you may "lose" an activity that should be part of the to-be process. The map also gives you a quick reference as to which work units and people will be affected by the reengineering effort. Also, without the proper level of detail you may lose sight of the points of intersection, where the process meets other processes that provide input into the process being reengineered. Finally, knowing the resource utilization of the as-is process helps in doing cost/benefit analysis of alternative reengineering options.

You usually don't need to map to the most minute level of detail (such as fax sent, fax received), but rather should stop at least one step above that level—the activity or transaction level (for example, record customer order). It is important to find the level of activity or transaction at which work is actually done and map to that level. When determining the level of detail to strive for, you need to balance the level of detail against the usefulness of the analysis for the effort being undertaken.

The complexity of the process being mapped may be instrumental in your decision as to what kind of process mapping technique to use—from a simple process flow diagram to complex, three-dimensional computer-based process modeling. Because the goal of reengineering is to simplify processes, we believe that you should always be able to describe the to-be process by means of a rather simple process flow diagram; if you cannot, you have not reengineered rigorously enough.

Figure 5-4 shows how you can create a set of process maps that "roll up."

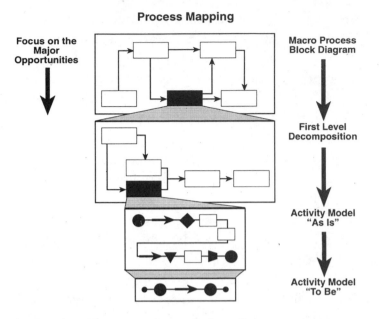

Process Mapping

Focus on the Major Opportunities

Macro Process Block Diagram

First Level Decomposition

Activity Model "As Is"

Activity Model "To Be"

Figure 5-4. *Creating process maps that roll up.*

A detailed activity analysis needs to be undertaken here. Since manufacturing processes have long been subject to industrial engineering, it is often easy to obtain activity-level information on them. This is less true for service and white collar processes: They tend to grow, spread, and mutate without having their work flows documented. In these latter cases, you may need to use an approach such as activity-based costing (ABC) or activity analysis to identify all relevant operations and the resources devoted to them.

By-the-book activity-based costing may be too time-consuming in a reengineering effort. We sometimes use an abbreviated ABC approach that locates all activities and generates approximate resource levels devoted to them.

In the data-gathering phase, you first identify the departments and work units involved in a business process. Through interviews with managers, you locate the activities that make up the process. Then you ask people who work on the activities to estimate the percentage of a normal work year that they spend on these tasks. (You can verify this, but usually it is on target.) Since service and white collar operations are labor-intensive, multiplying these time percentages by annual salaries is a decent approximation of cost.

There will always be differences of opinion about the as-is state. There will be those who say, "If only we could do the current process right, everything would be fine." There is often finger pointing by team members at other team members from other functions and disciplines.

It is important when doing your data gathering to *get the facts*. You need to conduct a fact-based analysis of the issues at hand:

When is the contract received?

When is it entered into the system?

When is the shop floor order released?

Who receives the change of beneficiary?

How does it get entered into the system?

When is the notification sent out?

One European telecommunications company said of its data gathering of the as-is, "We analyzed the way the order was processed and found that within one process there were several 'silos.' The key was to break down those silos and amalgamate several steps into one function; this immediately opened up time savings as well as cost savings."

THE IMPORTANCE OF INTERNAL CONTROLS

An effective internal control system provides senior executives with reasonable assurance that its most valuable assets—the entity's franchise and reputation—are well protected. Fallout from a publicly disclosed control breakdown can lead to a prolonged or even permanent impairment of reputation and shareholder value.

Legislators, industry regulators, customers, and the public are increasingly demanding that management be held responsible for developing and maintaining effective controls.

As you map the as-is process, you need to look for the internal controls in the process, and make sure they remain in the reengineered process. If you don't, you could end up like the bank that created a 15-minute mortgage that garnered increasing market share but unfortunately attracted a lot of bad credit risks. To achieve approval in 15 minutes, the company had removed many traditional controls from the process, and realized the danger in this only a few years later when a large number of the 15-minute-mortgage holders began defaulting.

You need to ask yourself the following questions about your:

Control environment. Is your control environment appropriate given the business processes you are focusing on?

Risk assessment. Are mechanisms in place to identify and respond to risks arising from external and internal sources?

Control activities. Do policies and procedures exist to control and carry out core processes?

Information/communication. Is there a process to identify, obtain, and disseminate internal and external information key to achieving your business objectives?

Monitoring. Do you have evaluation procedures to assess the adequacy and performance of your controls?

BEST PRACTICE 14: CHOOSE AND USE THE RIGHT METRICS.

Designing the process right means designing a set of metrics to use during implementation to make sure that the process continues to work the way it should. If the process doesn't have the right metrics tied to it, you'll never know if it's achieving the process vision.

Eighty-eight percent of our survey respondents agreed that using the right metrics is the key tactic in the ongoing monitoring of BPR implementation.

These metrics measure process performance, can be captured in the management information system (which may itself be transformed through BPR), and influence the correct behavior among employees who work in the processes being reengineered. There must be a logical interconnection between measures taken on the shop floor and those taken at intermediate steps all the way to the corporate measures that drive the business.

Process measures should include factors such as quality, service, time, and cost. Any functional measures that remain should not be in conflict with these process measures.

The difference between process metrics and traditional functional metrics is simple but profound. *Traditional metrics* were based on a model of input → activity at a functional level carried out by an individual → output. *Process metrics* focus on the output of teams and team members, on the way team activities and process performance enhances customer satisfaction and corporate competitiveness through improvement in quality, time, service, and cost.

Figure 5-5 shows the basic equation of process measurement. If quality and service are increasing while cost and time are decreasing, customer satisfaction and your competitive position should be getting better.

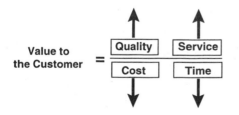

Figure 5-5. *The process measurement equation.*

Fully 67 percent of our survey respondents say that they put effort into creating the appropriate metrics, and are using those metrics to monitor reengineering progress during the implementation phase. Eighty-eight percent say establishing and using appropriate metrics are key to moving from the reengineering phase to a successful implementation. Figure 5-6 illustrates typical process metrics.

As one survey respondent from a footware manufacturer put it, "[We are] redesigning the formal performance measurement process *now* to insure that key metrics are in place prior to implementation of the reengineered process."

And seventy-four percent say they have used performance metrics to change and adjust individual performance goals. This is a key way to instill change in the organization.

Typical Process Metrics	
First Run Yield	The percentage of units that go through the entire process perfectly the first time, without delay, rework, or rejection
Cycle Time	The amount of time for a unit to complete the process, or subprocess from start to finish
Process Cost	Costs of activities within the process or subprocess, managed as a total system cost
Installation on First Call	When installation of equipment is completed correctly on first try
On Time Delivery	Percentage of total orders delivered to customers at the expected delivery time
Perfect Order	The percentage of total number of orders when customer receives exactly what is ordered, in perfect condition, at the expected delivery time

Figure 5-6. *Typical process metrics.*

Going from Goals to Measurement at Premier Bank

Sometimes the corporate vision and the stretch goals established at the start of the reengineering effort create the metrics that will be used to measure success. Take, for example, Premier Bank, a large commercial bank based in Louisiana.

Premier's top executives defined the corporate vision going into reengineering as being the:

> leaders in the financial services industry and the bank of choice in each market it serves. There will be no close second to Premier in terms of customer satisfaction, quality employees and financial performance. We will be universally acknowledged as the "premier" bank in Louisiana.

At the beginning of its reengineering effort, the bank looked to its three major core business processes: lending, taking deposits, and managing trusts. Bank executives decided that there were probably not enough opportunities for major breakthroughs in customer service or greater profitability in the reengineering of trust operations; so they focused on lending and taking deposits.

In the area of lending, the two major processes were selected for reengineering: commercial lending and consumer lending, both direct and indirect (indirect lending is so-called third-party lending through auto dealerships and the like). In the deposits area, the processes chosen for reengineering were new deposits and deposit transactions.

Premier established all its process measures from the perspective of customers. Across processes, the effort sought to:

Make core processes more responsive and timely.

Increase accuracy and reduce rework.

Get "closer to the customer" through enhanced delivery systems, sales and marketing, and organizational structure.

In reengineering the bank's lending operations, the team found that the key measure customers feel is important is cycle time: How quickly does the bank make a decision on my loan application?

Even if potential customers are ultimately denied a loan, they feel better dealing with a bank that comes to that decision quickly than they do dealing with one that dithers.

Bank executives were educated about the implications of cycle time and came to realize that, by focusing on cycle time and reducing nonvalue-adding steps in the process, they could accomplish the same goals they had sought to accomplish in the past with simple head count reductions. They could reduce the cost of the process and remove people from the process by removing unnecessary steps, and thus improve the profitability of the lending operation.

The steering committee set as stretch goals:

Reducing the time to act on a commercial loan application by more than 90 percent from the as-is condition.

Increase by nearly 70 percent the amount of loan officers' time that is actually spent with customers.

In the consumer loan area, the bank also set stretch goals: again to reduce time to act on the loan by more than 90 percent, as well as the explicit goal of reducing processing costs by 75 percent.

Goals and ultimately process metrics for new accounts were to reduce account-opening time by 65 percent, and to provide 24-hour access to accounts and to the means of opening new accounts.

In the deposit transaction process—including new account openings—a number of internal goals and metrics were established to meet the objective of increasing customer satisfaction to a score of "95." They included reducing paper by 70 percent and converting 50 percent of paper transactions to electronic transactions, reducing processing errors by 50 percent, and reducing expenses by 40 percent.

PROSPECTIVE BEST PRACTICE: CREATE AN ENVIRONMENT CONDUCIVE TO CREATIVITY AND INNOVATION.

The Holy Grail of a reengineering effort is the achievement of BreakPoints. *BreakPoints* are innovations you make in a core business process that are so profound as to cause disproportionate reactions in the marketplace. Federal Express's guarantee of delivery by 10:30 the next morning anywhere in the United States was a BreakPoint when it was conceived.

Of course, BreakPoints don't often last long. They become the industry standard, and the competition inevitably catches up. BreakPoints also don't happen in every BPR effort. The problem with BreakPoints is clear: How does a business actually identify them?

Some BreakPoints are self-evident. Others are relatively easy to spot through a rigorous analysis of current processes. Others drop out of a benchmarking activity. Sometimes, alerted to the possibility of their existence, innate experience and common sense identify them.

Two issues surrounding BreakPoints haunt senior executives. One is the feeling—rightly so—that, as a company and its competitors throughout the industry get better and better, it will become harder to identify BreakPoints. The second is that, if something is so obvious, how can it truly be a BreakPoint?

Such concerns are valid. BreakPoints are getting harder to find. And to find them deliberately—and in quantity—serendipity will not suffice. Properly approached, however, the probability of identifying and then achieving a BreakPoint is enhanced.

We believe that companies have to set out to discover BreakPoints—to identify, systematically and deliberately, new ideas and new ways of doing things. To do this, companies have to challenge the basic assumptions on which their businesses have been built—the "sacred cows" of the organization, of the industry, and of Western business.

Easier said than done, of course.

But it can be done, by focusing specifically on creativity and innovation. Many business executives are fearful of creativity and innovation. "We manufacture widgets," they say, "we are not media types." Or, "Our customers value our engineering ability, not our creative vision."

But increasingly we and others who work across a broad spectrum of clients in many industries and businesses believe that *the companies that survive and prosper in the 21st century will be those that foster the innate creativity in all their employees.*

Very simply, we define *creativity* as the generation of ideas and alternatives, and *innovation* as the transformation of those ideas and alternatives into useful applications that lead to change and improvement. We've found that, in today's business environment, an essential element to an organization's success is adaptability. You must be able to manage at the speed of change, and that takes creativity and innovation.

Creativity is not a management fad. Creativity is not an external system that can be imposed.

We all possess creativity. The ability to think creatively, to solve problems, and to invent new ways of solving problems is what makes us human. The management aspect of creativity comes about as businesses organize themselves to foster creativity. The hard work it takes to come up with creative ideas will be worthwhile only if a management system is in place that accepts ideas, tests and validates them, and gives recognition for those rejected as well as those ultimately accepted.

Although we all possess creativity, organizing that creativity is a learnable skill that can be enriched through practice, training, and new tools. A range of techniques can be used to promote creative thinking as teams organize and work together in a BPR effort. Word association, for instance. Or De Bono's six hats. Or even helping to break the mold of conventional thinking by postulating such deliberately provocative statements as, "Imagine that our company had no sales force."

Executives at Reebok, the footware company, posed just that kind of thought. They imagined a shoe manufacturer that didn't actually manufacture any shoes. And the company has created a very successful enterprise, designing and marketing athletic and casual footware, leaving the actual production of products to contract manufacturers.

Not all creative solutions are breakthroughs; many modify or improve on current products, services, or delivery systems. But for any improvement—breakthrough or incremental—to happen and be sustained over the long run, employees must challenge assumptions and create solutions.

Whatever the technique you use, the intention is the same: encouraging people from within the business itself to look at problems with a fresh perspective.

The stretch goals you establish at the outset of the BPR effort tend to force teams to be creative and innovative. But merely having the goals is not enough, the teams must have tools and techniques as well. As a part of team sessions during the reengineering phase, we suggest that teams spend some time on creativity training and on doing creativity exercises, both as individuals and as teams. There are a number of good creativity exercise workbooks.

Innovation takes creative thinking and turns it into practice. It moves the hard work of being creative in teams to the reality of the work process being reengineered. We find that many times a breakthrough in the team effort occurs when things simply seem to gel. Some call it the moment of collective, "Ah ha!" Often, it is the introduction of the value-added and nonvalue-added mapping techniques that leads people to this state.

But also, that collective ah-ha is when someone comes up with an idea that simply breaks the mold, by taking the old way of looking at the business and turning it on its head.

At Premier Bank, the mold was broken in the commercial lending process. The focus was correctly on the bank's customers, including the oft cited complaints of commercial customers that the bank's loan-approval process took too long and the loan officers seemed to obtain the necessary information in a haphazard, disjointed way.

In addition to reengineering the process and taking advantage of available technology to reduce the cycle time of the loan-approval process, and to achieve cost reductions through the elimination of nonvalue-added work, the bank decided to design "extra customer capacity" in the process, so that loan officers could spend more time with customers. The additional time, in conjunction with more streamlined information-gathering procedures, will pay huge dividends in terms of enhanced relationships and growth opportunities.

PROSPECTIVE BEST PRACTICE: TAKE ADVANTAGE OF MODELING AND SIMULATION TOOLS.

Computer-based process simulation, sometimes called computer-based process mapping, came into general usage in industrial applications—power plants, chemical plants, machine tooling—in the 1970s. Only in the mid- to late 1980s did they begin to appear for use in paper process flows, the so-called "back-office" operations in financial services, insurance, and other industries.

In the back office, where a great deal of process simulation for reengineering efforts takes place, building a model requires four types of data:

1. Process flow data (activities and tasks).

2. Resource data (people, machines, support systems, even whole business units).

3. Input data (product mix, demands, and volumes).

4. Event data (when events occur such as meetings, accounts closing, etc.).

From this, calculations can be made about current operations and future possibilities using quantifiable metrics such as the number of transactions, unit costs, costs of rework, resource utilization, and cycle time.

Process simulation works best in complex environments with high-volume systems that have alternative processing conditions. Simulation software models the as-is state through rigorous data gathering, and then can be used to test what-ifs when creating a vision for the to-be state. Until the advent of simulation software, it was hard to come up with baselines and visions for white-collar operations.

It is critical that data collection be done right the first time. It can't be an ad-hoc effort.

Simulation software also helps you calculate process costs—not only the total process cost, but also the cost of variance processing and the cost of individual activities within the process on a per-product or per-transaction basis. Often true costs are hidden within aggregated numbers.

Using simulation software is akin to using spreadsheets. And as with spreadsheets, there are two gains for users. One is in the data gathering effort itself. Filling in the structure to create the as-is model causes management to make decisions about job categories and other issues that may never have been explicitly stated. The second is in the ability to do numerous analyses and evaluations of alternative courses of action.

At the end of 1991, the information research firm IDC predicted that the market for dynamic business modeling tools would increase from $10 million (1990 market) to $150 million by 1996. It further predicted that such software modeling tools would continue to be proprietary, and be used by BPR consultants as part of their work.

CASE STUDY

Aetna Simulates Life Insurance Division Customer Service

In 1991, Aetna concluded that its Life strategic business unit would have to be reengineered. The unit was operating in a tumultuous business climate. Executives were forced to rethink the company's customer focus at the same time they were searching for ways to lower the cost structure. They were facing the classic dilemma of figuring out how to do more for customers with fewer resources, especially people.

One of the key precepts of the reengineering model called for a service operation built on the idea of centralized service, augmented by an 800 number.

Customer service in life insurance was considered to be both service to policy holders and service to agents. Services included processing changes in beneficiary, changes in address, loans against policies, or partial surrenders of cash value. In addition, there were statements for taxes (dividend income) and premiums. The same customer service people processed premium payments.

The organization at the time was split between field offices and the company's corporate headquarters in Hartford. The goal was to bring all the customer service work to Hartford and to create "islands of expertise"—specialty areas to handle the 20 percent of all customer requests that could not be serviced by generalist customer service reps.

The total customer service organization in Hartford numbered about 120 people. They were told that resources would not be added, even though they would be picking up work previously done in the field.

A Customer Service and Communication (CS&C) reengineering implementation team was assembled and given the responsibility of addressing the organizational issues of human resources, technology, work flow, and communication.

This group set about creating a new structure for operations; members used brainstorming, customer focus groups, benchmarking visits to "best of breed" customer service organizations, and finally work flow analysis. To accomplish this analysis, the team used sophisticated computer-based work flow simulation software.

Aetna took representatives from across functions and for several weeks gathered comprehensive data about how they did their jobs and how they related to one another. All the various processes that customer service reps were expected to go through—from a simple change of address to a more complex partial surrender of cash value—were mapped. It was eye-opening to see how many steps even the simplest process takes. In some instances, there were more than 1000 steps in a single process.

In the to-be state, all the various functions would be collapsed into generalist customer service reps, who would be expected to complete 80 percent of service requests, with the other 20 percent, representing the most complex processes, assigned to the islands of expertise. This meant that multiple processes would be handled by people with the same job description.

Clearly, one could describe from this data "what makes a good customer-service rep" in terms of knowledge, skills, and personality. From there, hiring criteria and training programs could be developed.

What-if scenarios were developed. What if the volume of calls to the 800 number varied by plus or minus 100 from the average on any given day? Plus or minus 200? What if the volume of complex calls increased against the volume of simple calls. Where might backlogs occur? How could resources be moved to respond?

Organizational Gains from Simulation

In addition to the technical abilities of computer-based simulation tools, they also have organizational advantages. The best tools, and the best facilitators/consultants who help companies use the tools, get the simulation done using the knowledge of the people there. People who actually perform the tasks validate and verify the model as it is being built.

At Aetna, simulation helped increase employees *and management's* buy-in of reengineering, because the "experts" had defined what they do and created ways they could do it better. The team process of getting data for the model leads to better communication among the groups and a better appreciation of what different departments or people do.

We have found that, when we have computer-based simulation, very often the key to worker acceptance of the findings is quantifiability. It's one thing to say costs are high because you have to clean up after people by performing rework. But it's another to motivate employees to perform so that rework is not needed in the first place. That can only happen if you can pinpoint the impact of specific tasks and activities that need improvement.

Another way that computer-based simulation enhances employee buy-in and future actions is by creating a common language. It is imperative that you create up-front agreement among the data-gathering team on the terms in describing activities and the characteristics of process performance.

MAKING BPR STICK

Best Practice 15: Understand the risks and develop contingency plans.

Best Practice 16: Have plans for continuous improvement.

Prospective Best Practice: Align the infrastructure.

Prospective Best Practice: Position IT as an enabler, even if the extent of the IT change necessary is great.

I f your company has done a good job in the planning and designing phases of the reengineering effort, the implementation should go relatively smoothly. However, since the reengineered process often fundamentally alters the way people work, you must take care to maintain active change management and communications.

Remember, implementation is different for every company engaged in a BPR effort; it is even different for each process that is reengineered.

Activities during implementation run in five parallel streams, which together create a top-to-bottom transformation of the business' operations. These five streams are:

Project management.

Process transformation.

Information systems.

Change management/organization.

Key performance measures and other procedures.

During this period of transition and transformation, you may:

Hire new staff, or transfer or retrain current staff.

Open new facilities, close old facilities, or redesign the physical layout and flow of exisiting facilities.

Purchase or develop new information systems, or enhance current systems.

Implement new operating procedures.

Accomplish a host of other activities that make up the total reengineering effort.

The underlying principles of such an implementation are well known and have been proven in organizational management:

Decisions must be fact-based.

Decisions should be made as often as possible by the people who perform the work, who know it best.

Teams are more powerful than individuals.

It is a good idea at this point to organize implementation teams made up of people who will be working in the to-be process. The more people you can include at this point the better, since this will encourage acceptance. But at least some of the team members should be carry-overs from earlier phases of work, since team continuity is important.

Planning needs to be driven to another level of detail, including determining resource requirements, technical fine-tuning, communications with other processes, and all the other details of putting a new process or system into operation. Also, it is wise to confirm the cost-benefit analysis done earlier and to revisit the issue of performance measures.

Process owners and implementation teams need to have the actual transition to the to-be process well choreographed. However, complex changes, even when well planned out, tend to run into a few glitches, both during installation and immediately after. Also, other processes that are connected with the new ones may require more minor modifications or enhancements to adjust to the new performance level.

It is important that people start using the process as intended, not running it in parallel with the old process. There must be a clean break, and specifications for the new process must be met to prevent "sloppy" habits from becoming embedded in the new process.

Formal training, with close attention by management, for several months after the new process is initiated is the best way to assure that all participants know how to use the process and what is expected of them.

Few of our survey respondents, only 2 percent, did not feel they had done "extremely well" during implementation. But 58 percent felt they had done better than "well," and another 40 percent felt they had done well. However, all participants felt that doing well in the implementation is extremely important, and many felt this is where they could improve on future efforts. By and large, they felt they did better in the reengineering phase than in the actual implementation.

The best practices in implementation have to do both with recommitting the organization to the task at hand, through reaffirming a willingness to accept risk and to plan for it, as well as with positioning the organization for the future beyond the excitement of a grand implementation. Setting up continuous improvement plans, setting performance measurements that reflect the reality of the new process, positioning the IT system as a partner and enabler for the new processes, and most importantly aligning the human resources infrastructure of the company are ways to increase the chances that the changes will be long lasting and effective.

BEST PRACTICE 15: UNDERSTAND THE RISKS AND DEVELOP CONTINGENCY PLANS.

Only 44 percent of our survey respondents said they were willing to accept more than a modest amount of risk in the implementation.

There are two types of risk: technical risk and organizational risk.

One technical risk is the risk that you will reengineer a core business process and implement the reengineered process, and it will not work as intended. Another technical risk is that, while you are implementing the reengineered process, ongoing business will be disrupted to such an extent that you will have deteriorating relationships with key customers.

By far the greatest organizational risk to a BPR effort is *cultural pushback.* By this we mean a reaction against the changes as being antithetical to the corporate culture. Even if people have become aware of the effort, and even if they have engaged in communication during the planning and reengineering phases, when the actual implementation begins and the changes hit home, they can be unnerving.

One respondent, from a major U.S. forest products company, describes risk this way:

> The initial perceived risk was that we can't conceive of something truly breakthrough enough to warrant this kind of major change. The perceived risk five months into the project was, "how can we assimilate this much change without disrupting the customer?" This was not apparent up front.

Sometimes the cultural pushback is so strong that it demands strong action by leadership.

A respondent from a major U.S. white goods manufacturer says,

Even when the change is all win there is resistance due to human nature. In some problem cases, the CEO took the reluctant group "out behind the barn" to have a heart-to-heart talk about the importance of the effort. Hindering improvement was not an alternative.

One way to fight this cultural pushback is through a rededication to thorough, effective, and truthful communications. People will not remember what was said in earlier rounds of communication; you must say it again. These must take place on a regular basis with people outside the implementation teams.

Fifty-seven percent of the companies surveyed held kickoff meetings with managers at the beginning of the implementation. In addition, 49 percent created companywide newsletters for the implementation period; a further 49 percent had question-and-answer sessions for employees; and 38 percent held workshops. Thirty percent sent individual letters and other communications to employees. Figure 6-1 shows how many companies used these various types of communications.

Despite this, 37 percent of respondents cited "not enough communication with employees" as one of the major failings of their implementation.

The message in the communication also needs to change. At the credit arm of a U.S. computer manufacturer, a respondent says, "They learned that major change never ends. Employees listened to every word management uttered, and the message changed from 'this is the answer' to 'we are never done.'"

About half the companies surveyed made a formal presentation to senior executives before moving from the reengineering phase to implementation. By getting an explicit "sign-off" to go ahead, they cemented commitment from the top. Communication of that kind of commitment throughout the organization can do a lot of temper resistance.

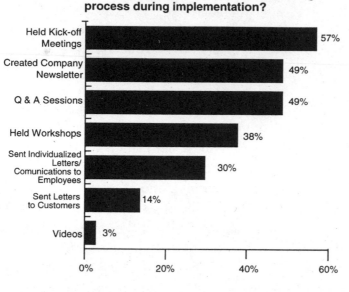

How did you communicate the details of the change process during implementation?

Method	Percentage
Held Kick-off Meetings	57%
Created Company Newsletter	49%
Q & A Sessions	49%
Held Workshops	38%
Sent Individualized Letters/ Comunications to Employees	30%
Sent Letters to Customers	14%
Videos	3%

Note: Multiple responses permitted.

Figure 6-1. *Companies using various types of communications.*

As well, 78 percent of the companies surveyed used retraining as a method to prepare the organization for the implementation.

One of our prospective best practices, aligning the infrastructure (discussed later in this chapter), is specifically aimed at reducing the risk of cultural pushback.

The best way to mitigate risks—both technical and organizational—is to implement reengineering through pilots that demonstrate success. Remember, when you are talking about enhancing relationships with customers and a new way of doing business, you need to pilot and demonstrate success.

This sounds obvious and even easy, but it is really frought with challenge, especially if you do it right. Doing it right means not implementing a subprocess, but an entire process on a small scale. Remember, this is not proving the concept through prototyping or simulating; this is actually making the reengineered process work for the customer.

An example of this kind of piloting comes from the reengineering effort of one public education district. One of the top ten school systems in the county, both in size and student achievement, this school system wanted to continue to be at the forefront of primary and secondary education.

As a result, four processes within the system were reengineered. Two have an impact on the quality of personnel in the classroom:

1. Reengineering the hiring process to ensure that top candidates are recruited and hired into the system.

2. Reengineering the staff development process to ensure that school personnel continuously upgrade their qualifications so as to remain technically and professionally current.

3. Reengineering system decisions. In the past, these were often made in an ad-hoc, "lone ranger," funds-available manner, which left the school system with disparate equipment, little technical support, and no interconnectivity. Now these decisions are linked to strategic business goals.

4. Reengineering the development of a progression plan for general education students, the forgotten "middle," as distinct from either special-needs students or gifted and talented students.

With 142,000 students and 198 schools, this process has an impact on almost every teacher and on most students in the system, and directly involves a key stakeholder group—parents—in the planning and monitoring of their children's education.

The individual and organizational benefits of the process are great, but so are the risks. The process requires five key transition planning conferences, with the individuals in the system who directly interact with students—teachers—planning for each student's progression. Figure 6-2 shows the timing and general goals of these planning conferences.

Process for progression planning conferences

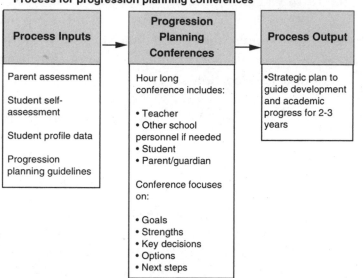

Process Inputs	Progression Planning Conferences	Process Output
Parent assessment Student self-assessment Student profile data Progression planning guidelines	Hour long conference includes: • Teacher • Other school personnel if needed • Student • Parent/guardian Conference focuses on: • Goals • Strengths • Key decisions • Options • Next steps	•Strategic plan to guide development and academic progress for 2-3 years

Schedule for progression planning conferences

After	Before
Kindergarten	Grade 1
Grade 3	Grade 4
Grade 6	Grade 7
Grade 8	Grade 9
Grade 10	Grade 11

Figure 6-2. *Timing and goals of planning conferences.*

A study of how teachers in the system spend their time showed that teachers on average work 59 hours per week, but spend only 1.4 hours per week in contact with parents. Implementing progression planning would represent a radical culture change for most teachers, who are not typically trained or prepared to work closely with parents in designing progression plans for individual students. In addition, the time to conduct progression planning meetings is viewed by teachers as an added burden to an already busy week.

The reengineering work team developed innovative approaches to freeing up teacher time to focus on this very important activity. But to convince teachers that the new process provides direct benefits to the students and their families, and that it assists teachers in doing a better job, its potential for success has to be demonstrated.

A full-blown pilot project will be undertaken in the spring semester of 1995, within one of the school system's "pyramids"— the combination of elementary schools that feed into middle schools, which in turn feed into a high school. This pilot will include all the computer support and student data necessary to implement student progression planning. It will also have in place a measurement component so the demonstrated results will be quantifiable.

While the pilot is in progress, the long-term implementation plan for the entire system will be moving ahead. Lessons learned and demonstrated successes from the pilot will be incorporated into the long-term implementation as the effort moves ahead.

BEST PRACTICE 16: HAVE PLANS FOR CONTINUOUS IMPROVEMENT.

An unmonitored process soon decays in performance. Without good performance measures and management attention, gains often unravel until you are once again fighting an uphill battle to become competitive.

There are two ways to address this need. One is to strictly maintain the status quo of the new process, holding onto the gains for as long as possible. The other is to encourage continous improvement.

Most people think of continuous improvement as small teams of employees who routinely enhance the operations they work in. However, that is only part of the approach. Actually, continuous improvement is shorthand for a style of *quality management* built around the goal of raising customer satisfaction through continuous process improvement.

Quality, as we use the term here, means not only the quality of products and services, but also productivity, efficiency, working environment, safety, ethics, corporate responsibility to the community, and every other value of any organization.

While the goal of BPR is to first find the *right proccess,* then determine how to *do the process right,* the goal of continuous improvement is to *do the process better.*

CONTINUOUS IMPROVEMENT AFTER A BREAKTHROUGH

BPR should always design in continuous improvement in order to sustain and improve breakthrough gains. Why are these small improvements so important? Here's what can happen to a breakthrough without them.

Performance can start to decay. Call it entropy, the relentless force in physics that pulls every ordered structure into chaos. It has many causes: People stop using standard procedures, equipment is poorly maintained, management loses interest after a breakthrough, and increased demands strain capacity.

Competitors achieve a new standard of excellence that makes yours obsolete—at least in the eyes of customers. This has the same effect on competitive advantage as the situations just described.

Factors change in ways that render your process incapable of performing as needed. These may include customer requirements, labor costs, energy costs, raw material availability, government regulation, or a host of other variables. The effect is the same.

If any of these things happen, you have two choices: Try BPR again or be second-rate. Both are expensive and possibly even demoralizing alternatives.

With continuous improvement, you gradually raise performance beyond the original standard of excellence set by the breakthrough. This means competitors will have to stretch harder and spend more money and time to exceed your constantly rising standard. Also, you use TQM methodologies to adjust the process to overcome, to the extent possible, factors that render it less than effective.

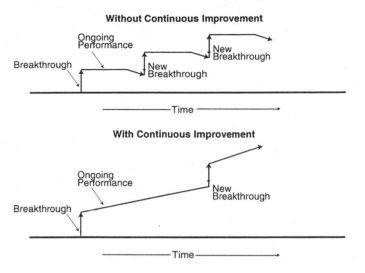

Performance After Breakthroughs

Without Continuous Improvement

With Continuous Improvement

Figure 6-3. *Result of the combined BPR/TQM strategy.*

The result of this combined BPR/TQM strategy is shown on the bottom half of Fig. 6-3, while the top half shows the effects of not following up BPR with a TQM effort.

Let's look at how one organization generated 40-percent improvement *after a breakthrough!* Remember the Pentagon's infamous $600 toilet seat? It belonged to the Navy's P-3C anti-submarine patrol aircraft. This story, about the same aircraft, takes place after the scandal and has exactly the opposite outcome.

C A S E   S T U D Y

Naval Aviation Depot's Breakthrough on the Navy's P-3C Contract

The Jacksonville, Florida Naval Aviation Depot faced a critical problem in 1988. It had just won an open competition with a private aerospace company to overhaul P-3Cs and refit them with new equipment. The losing firm told officials at Jacksonville, "You may have won, but you'll never be able to do the work at the price you quoted."

At first, it seemed this might be true, because the first few planes refitted came in way over budget. If this continued, the depot would lose the contract, and employees would lose their jobs.

To solve the immediate crisis, a cross-functional team of supervisors and employees completely reengineered the more than 100 task refitting processes, and eliminated many major time-wasting steps. This saved 500 labor hours per aircraft, a breakthrough success in labor-intensive work such as this. By the fifth plane completed, costs were back in line with the original bid price.

Most organizations would have stopped there. But depot executives moved ahead, reengineering the job's management structure with TQM, which resulted in saving 1200 additional labor hours off the total budget for the first group of planes. The new system worked as follows:

1. *Involving customers.* Previously, on-site customer representatives simply did quality checks, reviewing cost reports, and authorizing change orders. Now they were regularly asked ideas and made part of teams working on particularly difficult problems. Their input added new dimensions to creating improvements.

2. *Teams and ownership.* Before the contract, crews of single-skill employees did one part of the aircarft refit and worked on every plane. Thus, no employee worked on any one plane from start to finish.

For this job, employees were put into small multiskill teams, responsible for all work done on a single aircraft. This created a sense of ownership and pride for the final product.

3. *Rewards.* The ownership feeling was encouraged by having the teams engage in friendly and informal competitions to see who could develop the most improvements in productivity, efficiency, and quality. The rewards included t-shirts, caps, jackets, and even teaching their colleagues the new techniques. But most important was what one employee referred to as "the pat on the back you get."

4. *Delegation of authority.* Before the breakthrough, foremen and supervisors handled all the administrative duties of shop floor operations. In the new system, the team leaders, themselves employees, were given authority for tasks usually reserved for foremen: tool control, minor leave requests, and work assignments.

According to one foreman, "This is the first project in my 24 years here that I've actually been able to manage instead of just doing paperwork and putting out fires."

In the past, questions to customers had to go up the management ladder and down again to be answered, leading to delays and additional costs to work around problems that were not being answered. In the new system, foremen and team leaders were encouraged to talk directly with on-site customer representatives, and foremen were allowed to make some customer-authorized changes without higher clearance.

5. *Feedback.* On previous jobs only managers saw quality, schedule, cost, and material use reports, and it was usually weeks after the fact. In the P-3C job, performance data were quickly collected, analyzed, distributed, and discussed among managers, employees, and customers. The data was integrated into a computer graphic program of the work flows, which enabled speedy diagnosis of performance problems.

6. *Ideas from employees.* Improvement ideas once came mostly from managers. Under the new system, team leaders asked each member every week for suggestions on improvement. Since all workers were given a computer graphic printout of the work flow for their part of the process, they could point precisely to where improvements could be made. Team leaders came back with hundreds of practical ideas for speeding work, saving money, and improving quality.

Managers acted within days, sometimes hours, to implement these ideas. "We did this because we needed the improvements, not for morale," says one manager. "But what more positive feedback can you give an employee than to immediately implement a suggestion?"

The results: By the ninth aircraft, Jacksonville surpassed its best-case projection for labor hours, and by the 30th plane had trimmed another 1200 hours of the labor budget—40 percent less than originally planned.

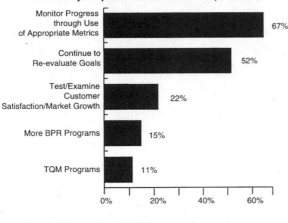

What are your plans for continuous improvement?

Monitor Progress through Use of Appropriate Metrics — 67%

Continue to Re-evaluate Goals — 52%

Test/Examine Customer Satisfaction/Market Growth — 22%

More BPR Programs — 15%

TQM Programs — 11%

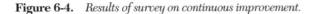

Note: Multiple responses permitted.

Figure 6-4. *Results of survey on continuous improvement.*

Many of our survey respondents said they have plans for continuous improvement, as seen in Fig. 6-4. Eleven percent have formal TQM programs, while 15 percent say more BPR efforts are anticipated. Twenty-two percent are closely monitoring customer satisfaction and market growth, and 52 percent are continuing to reevaluate goals.

At Motorola, a well-publicized adherent to TQM and BPR principles and a survey respondent, the quality drive will not end once Six Sigma is achieved. "Continuous improvement is an integral part of the program," the respondent says. "Motorola's new quality goal, restated in 1992, is to reduce defect levels by a factor of ten every two years from now to the forseeable future."

PROSPECTIVE BEST PRACTICE: ALIGN THE INFRASTRUCTURE.

To successfully cement reengineering changes into the fabric of the company's future operations, the company's infrastructure— its human resources, information systems, financial management, organizational structure, and support systems—all need to be aligned around the core business processes defined, as shown in Fig. 6-5.

We've spoken about human resources issues, most notably working in teams and focusing on team building and team skills (and measuring performance by teams and individuals acting in teams). Information systems make up the subject of our next prospective best practice.

Organizational structure (how reporting relationships reflect actual work flows) is an operational response to mission and strategy. Structural change tends to be meaningless without establishing and developing competencies of people in their new posts. If this is not done, old working practices will render the new organization little more than a rearrangement of the boxes, solid and dotted lines, and names.

Organizations must be viewed as dynamic, always evolving and reshaping to the demands of the business environment. For an organization to evolve continuously, it is necessary to allow freedom among the constituent units, to generate creative conflicts between them, and to encourage the application of new ideas.

INFRASTRUCTURE DEVELOPMENT

Figure 6-5. *Alignment of resources around core business processes.*

An appropriate metaphor for the modern organization is the "living organism." We often view organizations as "machines" that are in a static state until retooled by the top team. The living organism organization is based on creating self-organizing groups and has the following characteristics:

People at all levels think strategically.

Strategy is not regarded as the preserve of executives, rather it is the concern of everyone in any department at any level (the way the Toyota kanban system was developed). The systematic encouragement for creating strategy at all levels leads to innovative, mold-breaking thinking.

It is important that overall strategic vision should give broad direction, and allow freedom of interpretation at local levels. This freedom stimulates, empowers, and energizes people.

People focus outward by establishing networks with other organizations that have related technology and with customers.

Forming networks makes it possible to generate a variety of solutions and decision alternatives. This enriches the decision-making processes in an organization and raises the rate at which innovations occur.

These networks take many forms, such as joint ventures, personnel exchange, distribution networks, or acquisitions. General Motors and Toyota established the NUMMI joint venture at the GM plant in Freemont, California that made a spectacular success out of a history of failure, mainly because the organization was formed around self-organizing groups.

The hierarchy was abolished, the workers made the decisions that related to their work, and managers supported them. It was GM's highest producing plant and had one of its lowest levels of investment in technology.

Autonomous, multidisciplinary, goal-directed organizational units are formed.

Examples of this are small divisions, project teams, and task forces set up to take action and to generate creative conflict. Their success is measured by what they achieve and by what issues they uncover and address.

The emphasis is on business processes.

Organizations should seek to understand their business processes before seeking to redefine organizational structure. Most of the management controls necessary in a traditional hierarchical structure become unnecessary in an organization that has well-defined business processes, because such processes have built-in performance measures and so allow for self-monitoring.

The emphasis is on action.

Concrete action invites a concrete response. Therefore, action generates debate and movement. The maxim is: Act yourselves into a new way of thinking, not the other way around. To achieve this, the organization must support taking action and achieving results. The best way to do this is to form autonomous teams that are encouraged by senior executives to take action and get results.

People need support to survive and prosper in the living organism company. They need to understand and develop new competencies. A new organizational structure will achieve meaningful change only if the people with new jobs develop the capacity to do those jobs. So the extent to which the organization encourages people to learn new skills, knowledge, and attitudes is critical to successful restructuring.

People must become innovative and take risks. For people to become innovative and risk-taking, a learning atmosphere—or climate—needs to be developed in which experiences, whether successful or not, are rapidly assimilated and form the basis for learning how to cope with change.

When individuals are encouraged to learn, regardless of the immediate payback to the company in terms of productivity improvement, they begin to think differently. They challenge and look for answers to complex problems. In the long run, the rewards to the company for encouraging learning far outweigh the initial costs involved.

Continuous learning should be an integral part of every manager's job, and it comes from investing in training and development. This training should be integrated into the individual's career plan and accepted as part of the problem-solving process. These competencies must be seen as strategically important to reach the desired end state.

Training should be wide ranging to provide learning on the thinking and feeling levels, and be firmly geared to helping everyone in the company involved in the change process appreciate the impact of what they are doing.

By taking learning seriously, both on the individual and organizational level, a "living organism" organization is nurtured and able to renew itself. If learning is not taken seriously, the organization will drift back toward the bureaucratic hierarchical model.

PROSPECTIVE BEST PRACTICE: POSITION IT AS AN ENABLER, EVEN IF THE EXTENT OF THE IT CHANGE NECESSARY IS GREAT.

Much has been said about making sure to reengineer a process before adding the appropriate information technology, rather than simply laying technology over a cumbersome, inefficient process full of nonvalue-added steps.

Stephen Roach, the Morgan Stanley economist, has chided American business for spending far too much on technology that does just that. Too often, we believe, such actions have been driven by the traditional mind set of information technology professionals that they should find what their "internal customers"—the users of the system—want the system to do.

Instead, we believe the external customer should drive how information technology is used. The external customer should drive the needs of the people who interact with customers for information technology that helps them serve customers better.

Let's look briefly at how one company did just that.

*Carolina Power and Light: Better Information for
Better Customer Service*

The electric utility industry in the United States is poised on the brink of the type of dramatic change that the telecommunications industry went through in the 1980s. By the turn of the century, electric utilities will have moved from a heavily regulated, monopolistic, supplier-driven industry to a more competitive, customer-driven industry.

In the emerging environment municipalities and large industrial customers will increasingly be able to shop around for the most competitive price. In many cases, utility companies will no longer be restricted to operating in a particular region or area of the country.

Because energy is a commodity, the basis for competition will be price, just as it is in telecommunications, where the major carriers compete to introduce the most attractively priced packages to customers. Utility companies that may have only marginally focused on costs in the past will, in the future, be aggressively looking for cost-cutting measures such as reducing inventories by partnering with suppliers for Just-in-Time delivery of fuel.

But to even enter the price competition, utilities will need to meet some basic level of performance in two areas:

Reliability of service (consistent availability of power).

Customer service (accurate bills, timely changes in type of service received, etc.).

Information technology has a significant role to play in meeting the basic hurdles to performance in the utility industry, in terms of both reliability and customer service.

For example, systems linked to key suppliers can monitor the energy inventories and provide immediate information regarding delivery requirements to maintain uninterrupted service, while reducing inventory stores and nonvalue-added handling of energy-producing materials such as coal. Information technology can monitor performance of the delivery system to encourage predictive maintenance rather than either preventive or corrective maintenance. (Preventive maintenance is not cost-effective, since it is done prior to need, while corrective maintenance damages reliability, since the maintenance is performed only after an interruption in service.)

The level of customer service provided can also be greatly enhanced by information technology, and it was here that Carolina Power & Light (CP&L) chose to focus its initial BPR efforts.

Founded in 1908, CP&L provides electricity to more than 1 million customers in central, eastern, and western North Carolina, as well as central South Carolina. Headquartered in Raleigh, North Carolina, CP&L serves a 30,000-square-mile territory with a population or more than 3.5 million. CP&L power plants provide a flexible mix of fossil, nuclear, and hydroelectric resources, and its location facilitates the purchase and sale of power with other electric utilities, allowing CP&L to provide customers with reliable, cost-effective service.

In the mid-1980s, CP&L identified several corporate imperatives that would need to be fulfilled to ensure success in the new world order for electric utilities, which executives already could see over the horizon. The three top imperatives related to the "must-haves" in the utility industry:

Enhance CP&L's position as a leading provider of multiple-utility sources in its region (reliability).

Reduce, or at least hold down, present and future capital and operating costs (cost competitiveness).

Become more customer-focused (customer service).

By the end of the decade, CP&L had embarked on a corporatewide TQM effort, with a focus on reducing costs and improving customer service through business process improvements. CP&L recognized early that a number of breakthrough process improvements could only be made by replacing several aging core information systems, including the customer information system (CAIS). Top executives selected the replacement of CAIS with a new customer information management (CIM) system as the highest priority reengineering project.

Did the decision to replace technology drive or enable CP&L's BPR effort? This is like asking the old question about the chicken and the egg. The relationship between IT as a driver and IT as an enabler is not always black and white. The key is CP&L's approach.

We believe that the way CP&L structured the effort was clearly aimed at being customer-focused rather than IT-focused, and we think that enhanced the company's results.

To begin with, senior executives were clearly the owners of the BPR effort, not the information services department (ISD). Second, the project's sponsor was the senior vice president of customer and operating services. Third, reengineering work teams were co-led by one person from a business/functional department and one from ISD.

In this way, the effort was seen as customer-service-focused rather than IT-focused, even though CIM was the largest systems initiative ever undertaken by CP&L.

CP&L executives set several aggressive targets for BPR:

Support the corporate strategy to improve customer service.

Reduce net annual costs by at least $4 million.

Reduce cycle time in key business transactions.

Reduce field head count by 25 percent.

CP&L spent considerable time and resources learning about BPR practices, from a number of outside consultants as well as from companies that were further along in their own BPR efforts. The company also looked to experts and other companies for examples of successful major IT implementations.

The company combined this learning with its knowledge of quality and teamwork concepts, which had been honed through the corporate TQM effort.

CP&L used an outside consultant to provide an objective perspective on the utilities industry, for up-front work on IT strategy, and to provide an overall framework. But much of the project management work was carried out by a cadre of CP&L internal staff who had been recruited over the previous few years from Big Six and other management consulting firms to fulfill that role in a long-term reengineering effort.

During the course of the CIM project, CP&L developed both an internal Reengineering Council and a Reengineering Support Section, made up of the former consultants and others trained as project managers, to support the CIM undertaking and future BPR projects.

Through work flow analysis, process mapping, risk assessment, gap analysis, and vision sessions, the CP&L cross-functional senior management reengineering team performed an exhaustive analysis of the competitive issues and the processes proposed to be reengineered, as well as an examination of customer wants and needs. Their detailed assessment of the as-is helped them identify several major processes that would eventually have to be reengineered.

After meeting with the steering committee and senior management, the BPR team decided on the project parameters. Key concerns were resources, time constraints, head count, and, perhaps above all, which processes would have the greatest payback for the organization. From a practical perspective, the processes that could not be addressed through the CIM project were held for future BPR efforts.

It was estimated that one-third of the changes identified would need to be made in business policies or practices before changes in IT could be accomplished. Change management was actively undertaken. A change management team carefully planned for the utilization of human and capital resources.

It took the better part of a year to bring people from the various functions onto the reengineering work teams for the actual reenginering. The largest group of people who would use the new systems had special coaches to help them through the rough spots of the new system.

During the implementation, the company maintained contingency plans for the actual IT systems, but worked hard to mediate cultural pushback issues through extensive and exhaustive communications, training materials, and hands-on sessions. Despite the fact that it was the largest IT system change in 25 years—the entire working career of most users—most users reached the "comfortable with" to "power user" level of proficiency within six to nine months.

The project came in on budget, within one month of being on time, with all the planned business processes reengineered and IT changes made. More importantly, because business processes were changed, and change management issues actively addressed, the effort came in at a much lower cost than comparable IT systems projects within the utilities industry.

Results included:

Substantial cycle time reduction. (Figure 6-6, on the following page, shows the company's "read the meter this afternoon, bill tonight" cycle time.)

A flatter, leaner organization, with a reduction of approximately 125 positions, mostly through attrition.

Savings of $4 million annually.

Improved customer satisfaction.

In addition, the company has an improved ability to execute cross-functional change. Project success has served to build confidence in the employees' capabilities to work together to create major change, taking full advantage of the best that information technology has to offer.

Before CIM...

1. Updated meter reading to CAIS — Field Office

2. Error condition code received via paper report

3. Notify Field Office of error condition — Customer Accounting

4. Obtain re-read and notify Customer Accounting of results — Field Office

5. Receive re-read information and adjust TOU account

6. Cancel late payment charge

7. Calculate corrected bill using PC program

8. Manually produce corrected bill — Customer Accounting

9. Prepare corrected TOU bill attachment using PC program

10. Correct on-line TOU usage history

11. Mail corrected bill, attachment, and adjustment to Field Office

12. Review corrected bill, attachment, and billing adjustment

13. File billing adjustment — Field Office

14. Prepare customer letter explaining adjustment

15. Mail corrected bill, bill attachment, and letter to customer

Process time:	2 hours, 40 minutes
Total time:	Four days
Handoffs:	Four

With CIM...

1. Upload initial meter reading to CIM

2. Detection of reading error by CIM. Error condition reported immediately on-line to Field Office — Field Office Only

3. Issue on-line request to obtain re-read

4. Receive re-read results and do on-line adjustment

Process time:	30 minutes
Total time:	One day
Handoffs:	None

Figure 6-6. *Comparison of process characteristics, before and after reengineering.*

DEMING WAS RIGHT

It has become clear to us over the last few years, as we have worked with clients on Business Process Reengineering efforts, that the greatest success is achieved within companies that have already begun to move toward being customer-focused and market-driven in their external operations and process-focused and team-oriented in their internal ways of working.

If one thing is certain as we approach the 21st century, it is that change is going to be a constant in the business environment. Technological changes, including ever more rapid telecommunications, as well as the cost of capital and the ever changing demands of customers, are leading to such an environment. That means that a company has to have a constant improvement strategy.

As we enter the 21st century, success will be ever more difficult to achieve. The successful 21st century organization will be:

Fast.

Flexible.

Obsessed with continuous improvement.

To be sure, Deming was right in the way he originally consti-tuted Total Quality Management—large-scale improvement of the processes fundamental to success, combined with the contin-uous improvement of all processes. American businesspeople cre-ated the difference between TQM and BPR.

(When we say "Deming was right," we are really using shorthand to say that the whole panoply of early quality gurus were right, Deming, Juran, Feigenbaum, and Shewhart being the most prominent.)

Maybe the culture change necessary to work along both tracks simultaneously was too great for American businesses in the 1950s through the 1980s. Whatever the reason, it is clear that the most powerful results come by putting large-scale process improve-ment together with continuous improvement across the spectrum of activities, rather than keeping them separated.

We call a company able to do this an *Improvement-Driven Organization*.

> *Improvement* = change in the right direction.
>
> *Driven* = compelled to improve; there is no other choice.
>
> *Organization* = you and your employees, suppliers, cus-tomers, and stakeholders.

Improvement-driven organizations listen to customers, employees, and stakeholders. They focus on core business processes. They commit to both revolutionary and evolutionary change. And they manage all aspects of change: the technical, the human, and the organizational.

Figure 7-1 shows the dynamics of an improvement-driven organization. There are three drivers: the leader, processes, and improvement as a corporate objective.

The *leader* drives the improvement-driven organization by establishing the baseline: What is the company? Where is it going? What are its core business processes? The leader assigns the necessary resources to assess the current environment. Finally, the leader knows, understands, and communicates the vision and plan for the future, both near- and long-term.

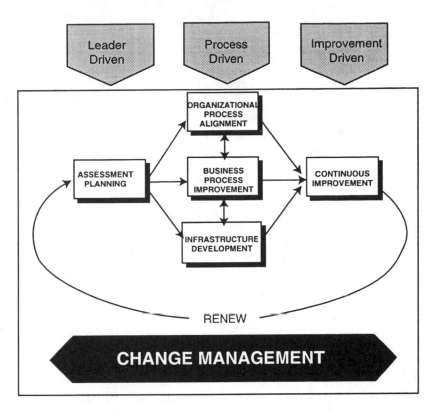

Figure 7-1. *Dynamics of an improvement-driven organization.*

Processes drive the improvement-driven organization because all processes need to be continuously improved; but all processes are not created equal. Some lend themselves to a quick fix, others need to be radically reengineered, and others can do with gradual improvement.

Yet process improvement without organizational alignment is like installing a new engine without changing the suspension. Sustainable improvement is not possible without fundamental changes in both the organizational culture and in the management processes and systems that make up the infrastructure supporting the culture. Hence, the need for *improvement as a corporate objective.*

Figure 7-2 shows how processes can be improved along the three improvement tracks.

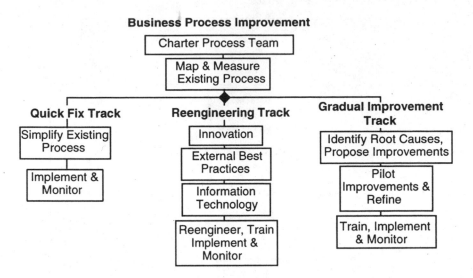

Figure 7-2. *How processes can be improved along the three improvement tracks.*

Sustaining the efforts of an improvement driven organization is not easy. It is leadership-dependent; the leader must continue to drive the effort. Sustaining and institutionalizing the changes are highly dependent on the degree to which the organizational culture has been aligned with process improvement, and how well the infrastructure that supports the organizational culture has been designed and implemented.

The effort is not truly institutionalized until the nomenclature of the improvement-driven organization is part of everyone's vocabulary and until the basic tenets are part of the way everyone goes about his or her daily work.

Becoming an improvement-driven organization requires leadership, patience, and hard work. There are no magic elixers. Now we'd like to tell you about one company that we believe has transformed itself. Over a period of about four years, it changed from an old-fashioned industrial giant with an entrenched culture into an improvement-driven organization that is fast, flexible, and anticipating the world it will have to live in as it enters the new competitive environment of the 21st century.

AlliedSignal: Total Quality Through Speed

This case study, written by T. Wood Parker, a managing associate in our Arlington, Virginia office, originally appeared in the Journal of Productivity Management.

The day Lawrence Bossidy became CEO of AlliedSignal, Inc., a company forecast predicted a negative cash flow of $435 million by the end of 1991 and $336 million in 1992, with debt 42 percent of capital. Executive morale was low.

In 1991, the market backdrop for the company's aerospace, automotive, and engineered materials products offered little comfort. The aerospace business faced a major decline in demand for U.S. military and commercial aircraft. The recession in Europe was cutting automotive sector sales. The slow-growth economies of both the United States and Europe made pricing increases difficult to obtain in all the company's business lines, shown in Fig. 7-3.

One year later, AlliedSignal had a positive cash flow of $255 million, an $831 million turnaround from the July 1991 forecast.

Aerospace Sector	Automotive Sector	Engineered Materials Sector
Propulsion engines	Brakes	Carpet
Turbine engines	Safety restraint systems	Industrial fibers
Avionics	Filters	Chemicals
Auxiliary power units	Spark plugs	Advanced materials
Environmental control systems		Circuitboard laminates
Wheels		Automotive catalysts
Brakes		
Guidance systems		

Figure 7-3. *AlliedSignal's lines of business.*

Sales went from $11.83 billion in 1991 to $12.04 billion in 1992. Earnings per share rose from −$2.00 to +$3.80. The company continued to make significant gains after that.

A major part of AlliedSignal's breakthough and continued success had been its strong focus on speed as a means of improving productivity and profits. In fact, speed itself is one of the company's seven corporate values. To put this value into practice, the company added a second component to the Total Quality initiative Bossidy started shortly after arriving. Total Quality Through Speed (TQS). (TQL began in early 1992, and TQS was instituted later that year and gained momentum in 1993.)

AlliedSignal's early TQ initiative, known as Total Quality Leadership (TQL), helped the company change its corporate culture and develop critical improvement skills. TQS built on this foundation of skills and training, explicity raised expectations for dramatic improvement in business results, and created a corporatewide focus on overhauling and reducing the cycle time of core business processes.

(As explained in Chap. 2, *TQS is what we refer to throughout this book as Business Process Reengineering*. Bossidy grasped that it was important to stick with the vocabulary of Total Quality, rather than shifting and calling the new effort something else. Thus he built his own momentum on the buy-in that the TQL effort had already achieved, and side-stepped the charge of bringing a "flavor of the month" mentality in his change efforts.)

The Building Blocks of Rapid Change

Bossidy instituted numerous organizational changes at AlliedSignal. He began by setting aggressive goals for performance on a variety of financial measures: He wanted:

A 6-percent annual gain in productivity.

An increase in operating profit from 4.7 percent in 1991 to 9 percent in 1994.

An increase in working capital turnover from 4.2 times per year to 5.2 times per year.

An increase in return on equity from 10.5 percent in 1991 to 18 percent by 1994.

Along with these goals, AlliedSignal's top 12 executives developed and communicated the company's to-be vision, shown in Fig. 7-4. Committing the company to "strive to be the best in the world," the vision specifies a customer focus, teamwork, innovation, and speed as key means to the end.

The vision also specifies that AlliedSignal will become "a Total Quality company by continuously improving all of our work processes to satisfy our internal and external customers." The definition of a Total Quality company includes:

1. Producing satisfied customers.

2. Focusing on continuous improvement.

3. Having highly motivated and well-trained employees.

According to Jim Sierk, AlliedSignal's senior vice president for quality and productivity, it also includes taking high-quality steps to achieve excellent business results, such as managing by fact, maintaining a process focus, and using a business-planning process that has quality goals, steps to achieve them, and the means to measure them.

Changing the Culture: TQ Training for Every Employee

Attaining the corporate vision meant changing fundamental aspects of the AlliedSignal culture and operations. Bossidy discerned that, to achieve the corporate vision, fundamental aspects of the AlliedSignal culture and managerial philosophy had to change.

For example, the managerial style was traditional and control-oriented. Decision making and information were tightly held. And the workforce was unfamiliar with team-oriented problem-solving tools. Bossidy, who had seen the value of Total Quality as vice chairman of General Electric, decided to use TQ at AlliedSignal to give employees the basic tools, framework, and common lexicon to improve work processes and to work successfully in teams.

The company hired outside consultants to work with senior managers to design an aggressive, hands-on Total Quality training program that would encompass every employee and directly

Our Vision	We will be one of the world's premier companies, distinctive and successful in everything we do.
Our Commitment	We will become a Total Quality Company by continuously improving all our work processes to satisfy our internal and external customers.
Our Values	**Customers** Our first priority is to satisfy customers.
	Integrity We are committed to the highest level of ethical conduct wherever we operate. We obey all laws, produce safe products, protect the environment, and are socially responsible.
	People We help our fellow employees improve their skills, encourage them to take risks, treat them fairly, and recognize their accomplishments, stimulating them to approach their jobs with passion and commitment.
	Teamwork We build trust and worldwide teamwork with open, candid communications up and down and across our organization. We share technologies and best practices, and team with our suppliers and customers.
	Speed We focus on speed for competitive advantage. We simplify processes and compress cycle times.
	Innovation We accept change as the rule, not the exception, and drive it by encouraging creativity and striving for technical leadership.
	Performance We encourage high expectations, set ambitious goals, and meet our financial and other commitments. We strive to be the best in the world.

Figure 7-4.   *AlliedSignal's "to-be" vision.*

improve business results. The TQL initiative began in early 1992, wtih consultants facilitating workshops for senior executives (including Bossidy), leadership of the three sectors, and all the strategic business unit (SBU) heads.

At the same time, the consultants trained a cadre of Allied-Signal master trainers to provide four-day training workshops to the rest of the workforce—more than 90,000 employees. The in-house trainers completed the job of educating all AlliedSignal employees by December 1993, training teams to use TQ tools and techniques on the real problems they face.

Employee surveys taken before and after the TQL training found that the Total Quality orientation had an important bene-fit beyond skills development. Employees now understand and largely accept the company's emphasis on world-class quality and continuous improvement. They also seem more satisfied under the new approach. In addition, managers not only understand and support the company's vision and emphasis on TQ, but they also use its approaches every day: working with flip charts in meetings, assigning themselves team roles, employing problem-solving tools, and measuring their progress in creating an empowering environment.

Human Resources Changes to Support the New Vision

While AlliedSignal executives were confident that the TQL initia-tive was providing excellent training, they recognized that the sys-tem driving behavior—the Human Resources system—needed to be aligned with the vision for change. One of the first steps they took was to revamp important aspects of the hiring, advancement, and rewards systems.

Bossidy himself took charge of adding depth to the executive team. Through a combination of new hires, promotions, and reas-signments, he brought more than 40 new leaders into the top ranks. He also commissioned teams, led by AlliedSignal's president and three executive vice presidents, to develop a plan for reengi-neering the company's process for college recruiting, hiring, train-ing and education, and career development. Benchmarking best practices and conducting focus groups with employees helped shape the new process.

In addition, the employee-appraisal process was aligned with the cultural changes being sought by ensuring that employees are evaluated on their TQ behaviors as well as meeting other objectives. Systems were also designed to support creating a deeper human resources pool companywide.

Workers now receive special training to improve skills in areas they identify during their annual career development meetings. They also define their career goals, accent skills they might not be using on their current job, and list other places in the company they would like to work. Management education is another important thrust, as is a new policy to rotate managers among businesses, which had occurred only rarely before.

Senior managers have also developed ways to give public recognition to team achievements. In the aerospace sector, for example, the TQL "Wall of Fame" spotlights team activities and triumphs in a centrally located exhibit.

TQS Planning Workshops: Strategy for a "Fast-Break Offense"

While TQL training was still underway, AlliedSignal's top executives recognized a need to "take the next step" by moving into a full-court-press BPR effort.

In addition to creating the cultural change to a TQ environment, TQL had produced process improvements. However, to achieve dramatic business results, it was clear that the scale of improvements needed to grow. Executives decided to undertake a companywide effort to reduce cycle time in core business processes—an effort they called Total Quality Through Speed in order to maintain the TQ nomenclature.

This approach met their requirements for a customer-focused, process-oriented initiative that could produce rapid results. Cycle time reduction, measuring time and emphasizing speed, was also a measurement-based idea that flowed naturally from the TQL foundation.

And it was a practical concept that employees "could get their hands around," says Donnee Ramelli, vice president for total quality for the engineered materials sector. "It was clear after the first TQS workshop that TQS had the potential to double the net income of most of our businesses in two to three years."

The company also decided to introduce the cycle-time reduction effort through training sessions that continued the successful approach used in TQL. However, while every employee had received TQL training, Total Quality Through Speed was conceived as a top-down program that would target core business processes to accelerate improved business results.

To get the reenginering initiative under way, TQS planning workshops were held at the corporate level, at the sector level, and in every SBU in each sector. At the corporate and sector levels, participants were the highest-level executives, including Bossidy. At the SBU level, each workshop included the general manager of the SBU and about 15 to 30 of his or her direct reports (such as executives for finance and administration; heads of manufacturing, distribution, and technology; heads of businesses; and key managers of the core business processes).

The two-day TQS planning workshops had four objectives:

1. Help managers understand what TQS entails and the importance of increasing speed in their operations.

2. Identify cycle time reduction goals and opportunities.

3. Select processes for cycle time reduction, and charter teams.

4. Develop a TQS deployment plan.

1. Help managers understand what TQS entails and the importance of increasing speed in their operation. "In the aerospace sector, the value of operating faster was poorly understood at first," says Mark Shimelonis, the sector's senior vice president for total quality.

> The pace of our product-development and production processes must be closely coordinated with our customers' processes. Their schedules had always dictated our schedules. If we speed up, managers wondered, wouldn't that just create "white space"?

At the workshop we came to the conclusion that increasing our speed in concert with our customers' needs could help us deliver more for equivalent customer dollars. That's a competitive advantage that could be critical in our shrinking market.

The workshop began by defining TQS and dispelling misconceptions about what it takes to increase speed. As Figs. 7-5 and 7-6 show, TQS does not increase speed by making employees do the same things faster or work longer hours. Instead, it gives them the tools to find and elminate waste and nonvalue-added activities.

Participants also learned why TQS would focus speed-enhancing efforts on core business processes. Core business processes begin with the customer, shareholder, or market need. They cut vertically across functional, geographic, business unit, and even company borders. And they end when the need is satisfied. These processes are the most critical, because they are the inherent basis for the company's competitive advantage.

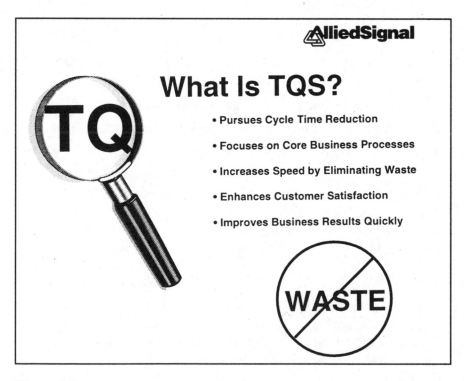

Figure 7-5.

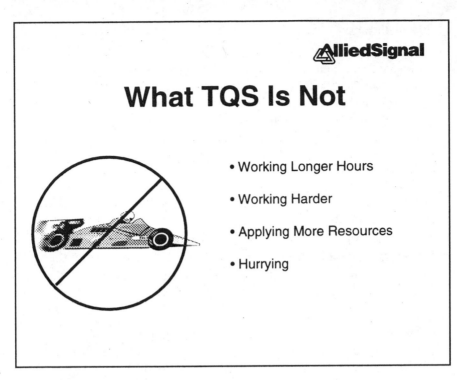

Figure 7-6.

Defining core business processes in an SBU was an important step in each reengineering planning workshop.

In the engineered materials and automotive sectors, for example, three processes stood out:

1. Customer-linked commercialization (taking a customer and market need through product development to production).

2. Customer-linked manufacturing (from customer acquisition through production and customer use).

3. Materials management (from sourcing through receipt and use of vendor materials).

Aerospace core business processes include new business development, order entry to collection, materials management, end user customer service, and integrated product delivery and support.

Ramelli says:

> Although this orientation was a new one for the company, managers really got behind the concept once they understood it. The idea of starting and ending a process with the customer was striking, and the people at the workshops came up with the unique terms "customer-linked" commercialization and manufacturing on their own.

2. *Identify cycle time reduction goals and opportunities.* Setting aggressive goals (stretch goals) for improvement was a critical part of each workshop. In the automotive sector, for instance, managers set an overall goal of cutting cycle time in half. In aerospace, the general goal was a 33-percent increase in speed.

Shimelonis says:

> One of the keys to our approach was to figure out our measures of success early and to use them. We chose outcome goals that related directly to improving aggregate productivity, such as how much overall improvement we wanted to see in each core business process, and how much of a process and a product would be covered by TQS initiatives. These broad outcome goals could both drive and measure the overall performance.

3. *Select processes for cycle time reducation, and charter teams.* Having managers relate time-reduction concepts to their real world, core business process operations was a central workshop activity. Participants brainstormed to identify aspects of their core business processes that could benefit from cycle time reduction, and in which it was feasible to increase speed through process improvement. They performed value-added analysis and drew "can-be" maps.

For example, managers concentrated on the parts of customer-linked manufacturing that cut cycle time in subprocesses such as product delivery, business development, order processing, production, and plant maintenance. Other managers began to look at the whole process from a reengineering or overhauling perspective.

According to Ramelli:

TQS maps and metrics identified a lot of waste and waiting time in our processes. This was crucial, because improving customer satisfaction, business results, and shareholder value depended on increasing speed and productivity. Not increasing the speed of our processes as part of improving them would have been playing with half a deck.

To get reengineering underway immediately, participants in each SBU planning workshop chartered TQS implementation teams to improve specific processes and subprocesses in its business. In the automotive sector, for example, workshop participants in each of the eight SBUs initially chartered two to four teams in each core business process. Over time, leaders added more teams, and even rechartered some teams to address cycle time issues. By late 1994, more than 250 teams across the company had worked on processes.

4. *Develop a TQS deployment plan.* During the planning workshops, managers created a plan for providing TQS training to staff who would be responsible for reducing cycle time in targeted core business processes. A critical aspect of each plan was its "bias for action"—an unmistakable push for rapid results in cycle time reduction.

Each TQS team working on a subprocess was expected to complete its work in 10 to 14 weeks. To help clear the path for action, managers identified potential barriers to success and brainstormed the actions necessary to overcome or mitigate them. This step was significant, because in many cases only the senior executives who participated in the workshops had the power to remove important barriers.

Participants in the planning workshops also created and staffed the infrastructure that would support TQS training, maintaining momentum in each core business process and in each individual team. The TQS deployment plans built on—and learned from—the approaches and organizations developed for the Total Quality Leadership effort. AlliedSignal again used consultants to "train the trainers," many of whom had conducted TQL workshops.

The TQS infrastructure was designed to provide visibility and support for TQS efforts at critical points in the change process. It also showed the entire workforce that some of the company's most senior executives were accountable for the success of TQS implementation.

As Fig. 7-7 shows, the support structure is comprehensive for each core business process, addressing long-term needs and facilitating interaction among SBUs.

TQS Infrastructure

Steering Group • corporate-wide • each sector • each SBU	Who: Top corporate leadership. Roles: Set goals; target processes for TQS; ensure training; identify and overcome barriers; monitor results; reward teams and managers, communicate about TQS.
Sector Process Leaders/Champions • one for each core business process in each sector	Who: Senior sector-level managers. Roles: Set goals and process objectives; advocate TQS; model management behavior; allocate resources; change policies to overcome barriers; communicate with steering group; monitor progress.
SBU Process Leaders • one for each core business process in each SBU	Who: SBU-level managers, "best and brightest." Roles: Work with sector-level Process Leaders to resolve problems; serve as focal point for day-to-day TQS activities; work with SBU Process Leaders; identify best practices and opportunities (SBU-level); build support for TQS in SBU; monitor progress; lead improvement efforts; identify barriers and communicate with sector process leaders and SBU Steering Group.
Team Champions • one for each team	Who: SBU managers who often have responsibilities in process being changed. Roles: Liaison with process leaders; communicate needs for resources or resolution of problems; report on progress; help develop team charter; review team products and process; guide development and measurement of performance indicators.
Team Facilitators • one for each team	Who: Come from all levels and functions; receive special training. Roles: Facilitate team meetings; assist in use of TQS tools; help develop reports on findings; work with the Team Champion to overcome barriers; get personnel and resources.

Figure 7-7. *A comprehensive support structure.*

*TQS Team Implementation Workshops: Training in TQS and
Application of Its Tools to Processes*

After the TQS planning workshops, TQS teams are "launched" by
working together at a TQS implementation workshop. These workshops include introducing team members to TQS concepts and
teaching the use of TQS tools and methodologies, such as process
mapping and value-added analysis.

Shimelonis reports:

> The key is having the teams learn these techniques by doing
> real work on their processes. Teams leave the workshop with a
> process map, barriers to speed identified, an action plan for
> improvement, and substantial progress toward achieving their
> objectives. They then go back to their worksites and look for the
> specific causes of nonvalue-added time in the process. They
> decide exactly how to reduce cycle time and develop a written
> plan for implementing the changes.

In some cases, teams found that the best way to speed up the
process was to reengineer it. In the automotive sector, for example,
a team designed a new, interactive process for prototype development that included working cooperatively with their customers,
Ford and Chrysler, as well as with AlliedSignal suppliers. They also
recommended adding new, computer-aided technology to speed
the design process.

In the aerospace sector, fluid systems increased speed in one
area of new business development by developing a *standard*
process. Following standardization, cycle time was reduced from
299 days to 90 days.

Teams in the engineered materials sector challenged every
aspect of their delivery process in an effort to make them faster.
One business switched from barges to trains to transport product,
reducing cycle time by 70 percent. The move also helped reduce
inventory by $1.7 million and generated a $375,000 sales increase.

In the automotive sector, converting serial activities to parallel
activities through better planning was another way teams met their
sectorwide goals of cutting cycle time in half.

Naturally, AlliedSignal's reengineering effort has faced—and continues to face—a variety of barriers to increasing speed. Some obstacles are human: the reluctance of people in any organization to do things differently and risk exposure of mistakes, as well as opposition to losing jobs through process streamlining. In these instances, the fast pace of the change itself has been an issue to deal with, as managers and employees have learned new skills, committed major energy, and adapted to a series of new situations in a short period of time.

Mark Shimelonis talks about the unique concerns of people who cooperate in an improvement effort that makes their own jobs expendable—because otherwise, in this market, even more jobs would be at risk. Disparities between the geographic dispersion of AlliedSignal companies have added to the complexity of creating major profit and cycle time improvements.

How has TQS succeeded? AlliedSignal leaders credit several critical factors with making breakthrough change happen by overcoming the obstacles. These critical success factors include:

1. Demonstrating vigorous leadership commitment.
2. Building from the Total Quality foundation.
3. Cascading TQS introduction.
4. Focusing exclusively on speed/cycle time.
5. Involving people with the power to make change happen.
6. Focusing goals and action on core business processes.
7. Providing consistent team facilitation.
8. Speeding up quickly.

1. *Demonstrating vigorous leadership commitment.* Lawrence Bossidy could "walk the talk." He consistently conveyed a sense of personal urgency about the need for change.

Shimelonis says management's personal involvement in the processes and monitoring of the monthly reports on progress sent the right messages about the seriousness of the improvement effort. Executives also showed it was important by linking TQS directly to the financial plan and the company's strategic goals. Other visible signs of support were in the breadth of the training effort and the resources applied, the open and ongoing communications program, and the everyday support the TQS infrastructure provided to those making the changes.

2. *Building from the Total Quality Foundation.* "Our workforce had already received Total Quality training, had developed a mature team effort, and was accustomed to using measurement tools," Shimelonis says. "These quality skills provided a critical foundation for TQS efforts, which added speed-related tools to our improvement toolbox. We couldn't have proceeded as quickly and as smoothly if we hadn't had that fundamental TQL experience."

In addition, TQL initiatives had made managers and the workforce aware of the importance of processes to productivity. This orientation helped TQS planners move logically to the more complex concept of core business processes. The quality maturity the company had developed through the TQL program also helped people plan and operate at the cross-functional level required for process engineering.

3. *Cascading TQS Introduction.* Planners believe that the top-down introduction of TQS was a major reason for its acceptance and success. As Fig. 7-8 shows, the cascade of planning workshops went from top corporate management to sector management, then to individual strategic business units. This allowed:

Top executives to define broad goals and strategies that helped direct activities at other management levels and achieve alignment across organizational boundaries.

Opportunities, barriers, and process definitions to be identified at a higher level and to become shared knowledge and the basis for common approaches.

Each level of management to appreciate the seriousness and commitment that this initiative was receiving at higher levels in the chain of command.

The Stages of TQS Planning and Implementation:
A Cascading Approach

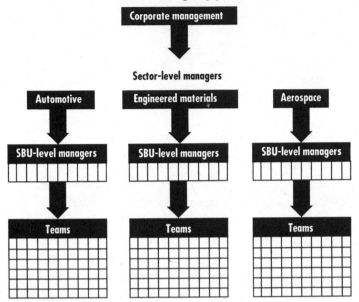

Figure 7-8. *Planning workshops.*

Lessons learned at each level to be applied in subsequent activities.

Modeling of management attitudes and behavior.

Figure 7-9 shows the relationship between the TQS planning workshops and the TQS implementation workshops.

4. *Focusing exclusively on speed/cycle time.* The diversity of AlliedSignal's business base required a flexible approach to improvement that would work in very different environments. Yet a shared goal was also desirable to provide a corporatewide focus for change and to make it possible to measure progress in a consistent, customer-focused way. Cycle time reduction allowed both, and it was also an easy concept for those involved to understand.

As one program trainer notes:

> Everyone knows how to measure time. In working with TQ teams, they always have problems deciding what to measure, how to know what to improve, and how to decide if it's worth

TQS Workshops

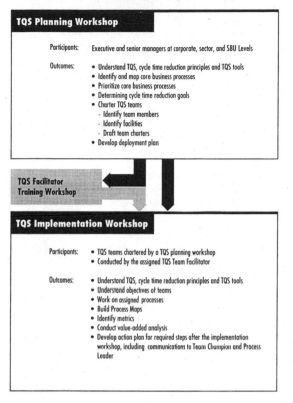

TQS Planning Workshop

Participants: Executive and senior managers at corporate, sector, and SBU Levels

Outcomes: • Understand TQS, cycle time reduction principles and TQS tools
• Identify and map core business processes
• Prioritize core business processes
• Determining cycle time reduction goals
• Charter TQS teams
  - Identify team members
  - Identify facilities
  - Draft team charters
• Develop deployment plan

TQS Facilitator Training Workshop

TQS Implementation Workshop

Participants: • TQS teams chartered by a TQS planning workshop
• Conducted by the assigned TQS Team Facilitator

Outcomes: • Understand TQS, cycle time reduction principles and TQS tools
• Understand objectives of teams
• Work on assigned processes
• Build Process Maps
• Identify metrics
• Conduct value-added analysis
• Develop action plan for required steps after the implementation workshop, including communications to Team Champion and Process Leader

Figure 7-9. *Relationship of TQS planning workshops to TQS implementation workshops.*

the effort to change certain aspects of the process. Focusing on time gets around these problems. Cycle time is relevant to almost every situation, and value-added analysis helps you focus quickly on waste and redundancy. When teams see that 10 percent of the time spent has value added and 90 percent does not, the need for improvement is obvious. You also find that resistance to the change goes away in the face of this clear, hard data.

5. *Involving people with the power to make change happen.* Although all employees participate in Total Quality initiatives, only selected managers and process owners have received TQS training and responsibilities—the pivotal people who have the power to reengineer processes and change policies.

"It wasn't feasible to involve the whole workforce in TQS," Shimelonis says. "It would have taken too long, and it wasn't necessary. Only about 20 percent of our people have the ability to change processes."

6. *Focusing goals and action on core business processes.* While core business processes differ among sectors, focusing on these horizontal, customer-oriented processes had increased the cohesiveness of the TQS effort and allowed meaningful measurement comparisons. Within sectors, strategic business units all work on the same core business processes, aligning efforts and creating synergies.

In addition, however, aiming cycle time reduction activities at core business processes puts the tools to work where they can do the most good. By definition, these processes are the most important activities of the business, the ones that generate value for customers. By speeding up these processes, significant outcomes can occur quickly, reinforcing the new ways of doing business and suggesting additional improvement opportunities.

7. *Providing consistent team facilitation.* During TQS, a cadre of facilitators provided support to different teams at different times. During evaluation, managers learned that a lack of facilitator continuity had been disruptive to group development and process-improvement projects. Profiting from this lesson, planners decided to assign one TQS team facilitator for the duration of a team's efforts.

Says Tom Hughes:

> Consistent, dedicated team facilitation helped our TQS teams learn to work together more quickly and move through the issues and disputes that all teams face. The speed and quality of process improvement efforts were positively affected as well.

8. *Speeding up quickly.* "It would have made no sense to pursue speed slowly," Hughes says.

> Our TQS process had to mirror our desired outcome just to be credible. In addition, AlliedSignal's reason for increasing pro-

ductivity was market-driven. The faster we improved our processes and our business results, the better our competitive position would be.

One key to rapid change was setting an aggressive timeline. Across the company, trainers conducted 94 training sessions around the world in about seven months. In the aerospace sector, for example, TQS planning workshops in fifteen locations took only about three months to complete. Train-the-trainer sessions followed a parallel track; once planning workshops had identified opportunities, and teams had been selected, TQS implementation training began right away.

"The workshops themselves also fostered rapid action," Shimelonis says. "If a team started at 8:00 A.M., by 10:00 A.M. they were already doing real work on their process; we wasted no time on academic exercises."

Another key was giving managers in charge a fast schedule for setting cycle time reduction goals—and sticking to it. Shimelonis recalls that:

> In the aerospace sector, managers were told in April to complete their specific goals and plans by September. Those who did were held up as models and rewarded; those who didn't were taken to task, and they had to hustle to catch up. Bossidy himself was monitoring this process, and his aim was to have us become a world-class company twice as fast as anyone else could.

Fast Breakthrough: Sustaining the Momentum

While AlliedSignal's fast break drive for excellence has already produced stunning results, the company has ambitious plans to sustain the momentum. One important initiative is providing TQS training to customers and suppliers—and creating teams that can reduce end-to-end cycle time in a process. In aerospace alone, about 50 joint AlliedSignal/customer teams are using TQS tools to address issues such as how to cut repair time on auxiliary engines.

"Our customers recognize that we've become a better supplier to them," Shimelonis says.

Our results have gotten us business, but they have also created customer interest in working together with us where processes overlap. We look for ways to take time out of the entire process and to improve the hand-offs between us and our customers.

On the supplier side, AlliedSignal has reduced the number of companies it buys from and set up long-term supplier contracts. It also has asked vendors to make aggressive cuts in prices and lead times—and offered them free TQS training to help. Pilot programs for supplier training have been completed, and training is now being made available to a wider group. Officials estimate that the new ways of working with suppliers saved AlliedSignal $100 million in 1993 alone.

Trainers note that suppliers in the pilot training were at first skeptical about the value of TQS and nervous about being able to meet AlliedSignal's cost and time requirements. By the second day of the workshops, however, they saw that no one was pointing fingers and everyone could come out a winner.

"The opportunities from value-added analysis made a real impression," one trainer says. "Suppliers saw that TQS was real; it could help them achieve the necessary reductions and improve their ability to compete."

AlliedSignal is also continuing internal efforts to improve business results and become a world-class competitor. New TQS teams continue to be chartered, with a particular emphasis on opportunities that can affect more than one SBU. The company also plans to provide additional training in quality improvement skills at regular intervals.

As Shimelonis says:

We want to become a learning organization that strives for customer-driven continuous improvement. Our success has given us the confidence to stretch further. Our fundamental skills in using teams, working in a disciplined process and measuring performance are creating a lasting capacity for positive change.

BEST PRACTICES

CHAPTER 2

1. Recognize and articulate an "extremely compelling" need to change.
2. Start with and maintain executive-level support.
3. Understand the organization's "readiness to change."
4. Communicate effectively to create buy-in. Then communicate more.

   Prospective: Instill in the organization a "readiness and commitment" to sustained change.

   Prospective: Stay actively involved.

CHAPTER 3

5. Create top-notch teams.
6. Use a structured framework.
7. Use consultants effectively.

   Prospective: Pay attention to what has worked.

Chapter 4

8. Link goals to corporate strategy.

9. Listen to the "voice of the customer."

10. Select the right processes for reengineering.

11. Maintain focus: Don't try to reengineer too many processes.

    Prospective: Create an explicit vision of each process to be reengineered.

Chapter 5

12. Maintain teams as the key vehicle for change.

13. Quickly come to an as-is understanding of the processes to be reengineered.

14. Choose and use the right metrics.

    Prospective: Create an environment conducive to creativity and innovation.

    Prospective: Take advantage of modeling and simulation tools.

Chapter 6

15. Understand the risks and develop contingency plans.

16. Have plans for continuous improvement.

    Prospective: Align the infrastructure.

    Prospective: Position IT as an enabler, even if the extent of the IT change necessary is great.

STUDY PARTICIPANTS

Aetna Life and Casualty
AlliedSignal
AT&T Global Information Services
Belgacom
Bell South
BFI
British Gas
British Telecom
Canadian Intellectual Property Office
Carolina Power and Light
Chevron Chemical Company
DHL
Digital Equipment Corporation
Eastman Chemical USA, Inc.
German Distribution Company
IBM Credit
ICL Retail
Income Security Department of Canada

Kooperative Forbundet

Lee Memorial Hospital

Marion Merrell Dow

Massachusetts Blue Cross/Blue Shield

Massachusetts Department of Revenue

Mitre Corporation

Motorola

Mutual of New York

Nike

Ontario Ministry of Finance

Philips

Polaroid

Rank Xerox

Raytheon

Reebok

Reuters

RF and Microwave and Electronics Company

San Diego Gas and Electric

SmithKline Beecham

Social Security Administration

Synoptics

Texas Instruments

UK Insurance Company

UK Electricity Generator

U.S. Postal Service

Volvo

Westinghouse

Weyerheuser Corporation

Xerox Corporation

BPR CONTACT DATA SHEET

BPR CONTACT DATA SHEET

[Interviewer will collect/confirm this information.]

Survey # _____

Primary business/SIC code _____

Company total revenues/assets _____

Major products/services _____

Location: Check ____ North America ____Europe

Name of organization/dept. _____

Primary contact person _____

Title: _____

Address _____

Phone _____

Fax _____

Other Contacts:

Name _____

Title _____

Address (if different from above)

Phone _____

Fax _____

Name _____

Title _____

Address (if different from above)

Phone _____

Fax _____

Name of operating company/organization (if different from that of contact person)

Date(s) of contact _____

Other Tracking information/notes regarding this interviewee:

_____

_____

_____

BPR TELEPHONE SURVEY— SCREENER AND PHASE I INTERVIEW

Note: Interviewer instructions in italics. "*" indicates questions of lower priority—ask if you have time.

First I'd like to ask you some brief questions about your BPR endeavor ...

Overall Project

1. When did you **begin** your current project?
   _____ (month/year)

2. At what **stage** are you right now? _____

3. When do you expect **completion**? _____
   (month/year)

4. Is this your company's **first BPR** or "major change" effort?

5. Can you tell me about your BPR project and which **process(es)** it involved?

6. Specifically, what target **goals** did you wish to accomplish?

7. Were the goals linked to your corporate strategy?

7a. If yes, how?

   [If target organization does not appear to be at least partially in the redesign phase, say,]

8. Thank you so much for sharing your valuable insights on your project. Would you mind if I called you back at a later date to get clarification, or more information, about your project?

Getting Started

9. What was the event(s) or **compelling factor(s)** that triggered the initial BPR effort?

10. On a scale of 1 to 5 (1 = not at all compelling, 5 = extremely compelling), how **compelling was the need** to change?

11. What was the **greatest risk** your company perceived in embarking on a BPR program?

***12.** What was the major difference between this BPR effort versus past efforts, if any, to change business practices or processes within your company?

The Role of Consultants

13. What role, if any, did **consultants** or other external parties play in your BPR effort?

[If no role, skip to Question 16.]

***13a.** If, applicable, what was the **role** and what is the **name** of the consulting organization or external party?

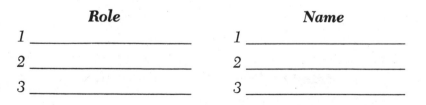

	Role		*Name*
1	_____	1	_____
2	_____	2	_____
3	_____	3	_____

14. On a scale of 1 to 5 (1 = not at all, 5 = extremely so), **how effective** were they?

15. On a scale of 1 to 5 (1 = not critical at all, 5 = extremely critical), **how critical was the role** consultants (or other external parties) played in the reengineering effort?

Teams

16. Did you **use** teams?

16a. If yes, How was the **initial project team** constructed?

16b. Who was on the team?

16c. Why were they **selected?**

16d. What was their **role** on the team?

Team Members?	*Why selected?*	*Role?*
1 _____	1 _____	1 _____
2 _____	2 _____	2 _____
3 _____	3 _____	3 _____
4 _____	4 _____	4 _____
5 _____	5 _____	5 _____
6 _____	6 _____	6 _____

17. How much **time,** on average, were team members **expected to devote** to BPR team activities e.g., 50 percent time, 30 percent time?

18. Did the team have a **charter?**

18a. If so, what was it?

19. Did the team have **special incentives and performance objectives** that were distinct from their usual (or day-to-day) job performance incentives or objectives?

19a. If so, describe them briefly.

20. Did the team receive **specific preparation or training,** such as in managing innovation, data analysis, simulation?

20a. If yes, What **type of training?**

*21. Did the **team change** over the course of the project?

*21a. If so, why?

*21b. How?

22. What **factors contributed** to the team's success?

*22a. And why?

23. And conversely, what **factors** did, or would, **inhibit its success?**

24. On a scale of 1 to 5 (1 = not well at all, 5 = extremely well), how would you rate the **overall performance** of the reengineering team?

25. On a scale of 1 to 5 (1 = not critical at all, 5 = extremely critical), how **critical was the team's role** in the reengineering effort?

26. If you had to pick **one thing** that the team did that was **most critical to the success** of the reengineering effort, what would that be?

The Role of Top Management

27. Who was the **primary sponsor(s)** of the BPR project?

28. On a scale of 1 to 5 (1 = not involved, 5 = extremely involved), how would you rate top management's overall **level of involvement** throughout the BPR effort?

29. On a scale of 1 to 5 (1 = performed poorly, 5 = performed exceedingly well), how would you rate the **overall performance** of top management in supporting and guiding the BPR effort?

30. On a scale of 1 to 5 (1 = not critical at all, 5 = extremely critical), how critical was **top management's role** in the reengineering effort?

31. What were the two or three most **important things that top management contributed** to the BPR effort?

*32. In hindsight, would it have helped to have other executives sponsor the BPR project?

*32a. If yes, who would that be?

Readiness for Change

33. On a scale of 1 to 5 (1 = not very resistent to change; 5 = very resistent to change), how would you rate your organization's overall outlook or **attitude toward change prior to** beginning the BPR effort?

33a. Why did you give yourself this rating?

34. Were there **other ongoing major change efforts,** such as TQM programs, at that time?

*34a. If yes, what were they?

34b. And were they **successful?**

35. Were there **specific communications** or events planned at the outset of the BPR program (to create the employee/shareholder/customer awareness or "buy-in")?

35a. If yes, what were they?

36. Were specific communications or events planned for execution **throughout the project?**

37. If you did plan/execute **specific communications**/events to enhance readiness, on a scale of 1 to 5 (1 = not at all, 5 = very effective), **how effective** were they?

38. On a scale of 1 to 5 (1 = not at all, 5 = extremely so), **how important is achieving organizational readiness** prior to initiating a PBR effort?

Selecting and Redesigning the Processes

39. Did you **use a structured or specific BPR methodology** in your approach in identifying and focusing on the areas for improvement?

40. Was the **"voice of the customer"** (or customer perspective) taken into account in the redesign?

40a. If yes, how?

Prioritizing and Focusing on the Process

Now I'd like to ask some questions about how you prioritized the processes to be reengineered ...

41. How did you **prioritize and focus** on which process(s) you would reengineer?

42. Looking back, on a scale of 1 to 5 (1 = not well at all, 5 = extremely well), how **well did you do in selecting** the process(es) for change?

43. On a scale of 1 to 5 (1 = not important at all, 5 = extremely important), **how important** was selecting the right process(es) to change to the overall BPR effort?

Starting Redesign

44. After you selected the process(es) to change, **how did you embark on the actual design?**

45. Did you start your redesign with the approach of gaining a comprehensive **knowledge of your process(es) as-is or the idea of "wiping the slate clean"** and starting over?

IT Component

46. Do you view **IT as an "enabler" or "driver"** of the reengineering project?

47. On a scale of 1 to 5 (1 = no or little change, 5 = massive change), how would you rate the extent of change that had to take place in your IT system, as part of the program?

48. What specific techniques did you use to redesign the process(es)?

48a. Which did you feel were most useful?

48b. Least useful?

48c. And why?

49. What was the **decision-making process for choosing/validating** the redesign plan?

50. What **role, if** any, did **external consultants** play in the redesign? (*If not answered in Question 13.*)

51. How did you **plan for the utilization of organizational resources** (probe for people and plant issues, such as staffing/displacement, outplacement budget, plant closings)?

52. Did you develop contingency plans?

53. What **risks and obstacles** did you anticipate **during this phase**?

53a. Which obstacles actually occurred?

Risks/Obstacles	*Occurred?*
1 _____	_____
2 _____	_____

54. Looking back, on a scale of 1 to 5 (1 = not well at all, 5 = extremely well), how would you **rate your overall performance in redesigning** the process(es)?

55. Again, on a scale of 1 to 5 (1 = not important at all, 5 = extremely so), **how important was the redesign** to the overall project results?

Transition to Implementation

56. What mechanism/decision-making process moved the organization from redesign to implementation?

Implementation/Realization

57. Specifically how did you **prepare the organization for implementing** the changes?

58. Specifically how did you communicate the details of the change process during implementation?

59. Were steps taken to make **adjustments and refinements** during the implementation?

60. What were the **major challenges** that occurred during this phase?

60a. How were they handled?

61. Again, on a scale of 1 to 5 (1 = no risk at all, 5 = a lot of risk), how would you rate the overall level of **risk that you were/are willing to accept during the implementation phase** in order to achieve your goals?

Education and Training

62. Throughout the implementation process, **how much employee training and education was required?**

63. Did the amount of **training and education** that employees received seem to be **adequate,** given their new job requirements?

64. Was **training and education** available/**provided** to the employees **as they needed it?**

65. On a scale of 1 to 5 (1 = not at all, 5 = extremely well), how would you rate the **overall performance during the implementation?**

66. On a scale of 1 to 5 (1 = not at all, 5 = extremely important), how would you rate the **importance of what you did in the implementation phase** to the ultimate success of the project?

Transition to Renewal/Monitoring Progress

*67. How did you know when you had reached the end of the project?

Monitoring Progress and Planning for Renewal

68. How did you **ensure the ongoing monitoring** of the implementation?

69. Did the company change/**continue to evaluate individual performance goals** and incentives?

69a. If yes, how?

70. Is there a **plan for continuous improvement** and further change **in the next few years?**

71. If yes, what **mechanisms** do you have in place **for making refinements/adjustments?**

Benefits/Impact on the Organization

*72. What was the **major difference** between this BPR effort and past major efforts to change business processes?

73. What **impact** did the reengineering project have **on the structure** of your organization?

74. On a scale of 1 to 5 (1 = totally dissatisfied, 5 = totally satisfied), how would you rate your **overall level of satisfaction with the progress, or results,** of the BPR program?

[This will depend on the stage the company is in at the time of the interview.]

75. On a scale of 1 to 5 (1 = not at all, 5 = very much so), **how realistic were your target goals?**

Assessment and Advice to Other Companies

76. What do you feel is the **one critical "must do"** (in order to achieve success) in a reengineering effort?

77. What do you feel is the **one critical "must not do"** (in order to avoid failure) in a reengineering effort?

78. Would you undertake this project again?

79. If you had to do **this BPR project** over again, **what would you do differently?**

*79a. And why?

80. What **words of advice** would you give to other companies undertaking a BPR effort?

*81. Were there any **_unanticipated benefits_** achieved from the reengineering program?

*81a. If yes, what were they?

Thank you for all your help. We will be sending you a copy of the executive summary of the report in six to eight weeks.

INDEX

Training, 159, 174
   and facilitators/consultants, 89-90
   teams, 72-73, 135-137
Transactional change, 37
Transformational change, 37

U.S. automotive industry, efforts to change, 34
U.S. Postal Service, case study, 56-63

Vision, refocusing, 132-133
Visionaries, 88
Voice of the customer, listening to, 115-119
   case study, 117-119
von Braun, Werner, 73

Welch, Jack, 41, 42, 45
Workshops, 72-73, 101-102
Work teams, *See* Reengineering work teams

Xerox Corporation, 39-40

Zalesnick, Abraham, 47

About the Authors

David K. Carr is a Partner with Coopers & Lybrand Consulting. He is responsible for the firm's Center of Excellence for Total Quality and Change Management. Since starting his consulting career in 1976, he has served a variety of private and public sector clients. He is the co-author of three books: *Excellence in Government—TQM in the 1990s, Breakpoint: Business Process Redesign,* and *Managing Change: Opening Organizational Horizons.*

Henry J. Johansson is a Partner with Coopers & Lybrand Consulting in the Information and Telecommunications Industry group. His career with Coopers & Lybrand began in 1976 with a focus on performance improvement for manufacturers, which led to his responsibility for Coopers & Lybrand's U.S. Manufacturing Industry practice. He has coauthored two books dealing with BPR, the latest entitled *Business Process Reengineering: Breakpoint Strategies for Market Dominance* (John Wiley, 1993).